GW00838887

The Merlin

The Merlin

Richard Sale

Snowfinch Publishing

To Susan, Without whom, Not.

First published in 2015 by
Snowfinch Publishing
The Beeches
Coberley Road
Coberley
Gloucestershire GL53 9QY
UK

ISBN 978 0 9571732 1 7

Acknowledgements

A book such as this cannot be completed without assistance and I am grateful to many people for help and inspiration. Nick Corley, Steve Davis and Mike Price helped in England, while Alan Heavisides and Graham Rebecca were of considerable help in Scotland. Of particular help in Scotland were Graham Anderson and Keith Burgoyne without whose unfailing assistance and consistent good humour the book would have been poorer and the field work more trying. In Iceland Jóhann Óli Hilmarsson, Ólafur Nielsen and Sindri Skúlason helped with both information and photographs, each provided with patience and unfailing good grace. I am grateful to Jens-Keld Jensen and Silas Olafson in the Faeroes, to Per Michelsen, Juha Hamkala and Torgeir Nygård in Scandinavia. In America Nick Dunlop, Tom Grey, Julian Hough, Laurie Keopke, Greg Lasley and John Warden were equally helpful. I am also grateful to Igor Dorogoi, Vladimir Ivanovski, Andrey Kovalenko, Eugene Potapov, Dasha Shergalina and Jevgeni Shergalin for assistance with data from across the vast Russian range of the Merlin. I am equally grateful for the hospitality and assistance of Gemma and Gerard Sulter who were willing to allow me to attach a unit measuring flight characteristics to their male Merlin (the mighty 'Atom') which provided valuable field information. I also thank the library services of Oxford University for their help and assistance, particularly with obtaining some of the more obscure of the reference material and the publishers of cited scientific journals for permission to use figures from specified articles. Finally, I thank my wife for her willingness to put up with my absences, and my bad humours when the weather and other circumstances threatened the project. As the dedication notes, this book would not have been possible without her unfailing enthusiasm.

Printed and bound in the UK by 4word Ltd, Bristol.

Contents

Introduction

For those who have journeyed close to the treeline in North America, Fennoscandia or Siberia, and stood with sparse trees to one side, Arctic tundra to the other; or have trekked through the volcanic landscape of northern Iceland where the view extends from jagged rock to the shores of a polar sea; or have hiked across a windswept high moor in Britain, one of many joys to be experienced is the sight of a small falcon, either high above or hugging the ground, travelling at speed, searching for its next meal. This is the Merlin.

The Merlin is a curious falcon. In a family of birds that divides reasonably comfortably into four groups it sits outside that organisation. In a family of birds which is, predominantly, found in warmer climes and richer environments, the Merlin stands out again, seeking harsher climates and less forgiving landscapes. Only the Gyrfalcon, the largest of all falcons, and the Peregrine Falcon, that inveterate wanderer, penetrate further into the Arctic. And only those two match the Merlin's circumpolar, northern range, the three species breeding in all the countries which define the Arctic Ocean, though Peregrines are absent from Iceland.

Merlins have long been a falconry bird, whereby individuals have been taken from the wild and bred in captivity, so allowing a basic knowledge of the species to be acquired. But for a true understanding, as with all species, a study of the bird in the wild is needed. The sheer size of the Merlin's breeding range might, at first, imply a species that is accessible and easy to study. But nowhere across its range is the bird common, and much of the land it inhabits is difficult, both to access and to work in. Recently Merlins have moved to urban areas of North America, which has made life easier for the researcher, but information on the species, though improving, is still scant. This book attempts to draw together what we know about this elusive, but fascinating, small falcon.

Opposite
Female Taiga Merlin, October, Connecticut, USA.
Julian Hough.

The Falcons

The origins of falconry are still debated, but it is certainly a sport which is several thousand years old and is likely to have originated in central Asia. That origin supports a minority opinion that the word *falcon* is of Germanic ancestry with tribes from areas now covered by Germany and eastern Europe learning their skills from further east. Majority opinion claims this is a false theory, instead, putting forward the idea that Germanic tribes learnt the sport from the Romans (though historic evidence does appear to suggest the reverse) who had acquired it from the east, the word *falcon* deriving from the Latin *falx* (the genitive of *falcis*) a curved blade or sickle, probably from the shape of the bird's talons or beak, perhaps even the shape of the wing. In old French *faucon* was the word for a diurnal raptor (*raptor* itself derives from the Latin *rapere*, to snatch or carry off) and as the French was the language of the English nobility after William's 1066 conquest, *faucon* became the old English form until around the 15th century when scholarly Englanders added an 'l' to conform to the assumed Latin origins. Other terms for falcons which also derive from the sport of falconry, and which are in (relatively) common use include *tiercel* (occasionally *tercel*) for a male falcon. This probably derives from the Latin *tertius*, meaning one-third, through the old French *terçuel* (occasionally *tiercelet*) because the male birds are, in general, one-third smaller than the females. Another term is *eyas* (occasionally *eyass*) for a nest-bound chick, which is assumed to derive from the Latin *nidus*, nest, through the old French *niais*, with *un niais*, being modified over time. Chicks are also occasionally referred to as *pulli,* though this is often used to refer to any chicks rather than just those of a falcon or raptor. Here the derivation is more certain, the Latin *pullus* meaning a young animal, though as it was also the Latin for chicken (the base of the French *poulet*, which became Middle English *polet* and hence *pullet* for a hen) there is room for negotiation, particularly as *pullulare* was the Latin verb 'to sprout' which could also be used to imply young growth.

The debate over the origins of *falcon* (and of falconry) is mirrored by those of falcon genera. The Family Falconidae of the Order Falconiformes is divided into two sub-families, the Polyborinae, which are largely confined to neotropical South America and include the caracaras and the forest falcons, and the Falconinae, which are more widespread and include the falconets and pygmy falcons as well as the 'true' falcons (of the genus *Falco*). As the behaviour of the caracaras resembles that of vultures and, to a lesser extent, buzzards, while the forest-falcons are hawk-like, this has led to the suggestion that there is a close evolutionary link between the Falconidae and the Accipitridae (hawk) families. However, that, and the suggestion of a South American origin for the Falconidae, are debated and will likely continue to be so into the future. Though fossil forms indicate that the two families have origins in the Eocene (some 35 million years ago) there are structural differences between the 'falcons' and the 'hawks', as well as differences in the moult sequence, though the often-quoted primary difference, that 'falcons' do not build nests whereas 'hawks' do, ignores the fact that caracaras do construct untidy piles of sticks.

Whether of South American origin – a suggestion supported by the number of genera and endemics found there – is true or not, the Falconidae do appear to have evolutionary roots in the Neotropics (Griffiths 1999) where many species are still found. Griffiths used a combination of morphological differences (specifically of the syrinx of various bird families) and genetics (specifically sequence based on mitochondrial cytochrome *b*) – see also Griffiths 1994a, 1994b and 1997 – to establish a view of falcon evolution. Griffiths (1994a, 1994b, 1999) also suggested that all diurnal raptors have evolved from a single ancestral family. Though the ancestral line of the falcons is still debated, particularly in the field of genetics, what is undoubtedly true is that the True Falcons are more closely related to the falconets (the Spot-winged Falconet (*Spiziapteryx circumcinctus*), the two Polihierax pygmy falcons and the Microhierax falconet species), than they are to the Daptrius, Milvago, Phalcoboenus and Polyborus caracaras, the Laughing Falcon (*Herpetotheres cachinnans*) and the Micrastur forest-falcons.

More recent molecular phylogenetic analyses (Hackett *et al.* 2008) imply a link between parrots and falcons at the ancestral level. Wright *et al.* 2008) noted a similar link, and suggested that an ancestral form of parrots and falcons originated on the supercontinent of Gondwanaland during the Cretaceous. The work of Pyle (2013) supports the close relationship between parrots and falcons, noting that the Falconiformes and Psittaciformes share a moult sequence which is not seen in any other order of birds. (One parrot, the Kakapo (*Strigops habroptila*) does not share the sequence, but Pyle considers it probable that this species split from the other parrots prior to the divergence between the Psittaciformes and the Falconiformes.) However, while this relationship has merit, genetics is an evolving subject (a somewhat ironic situation given its importance to evolution) and there are discrepancies between the results of genetical methods and data based on both paleo-environmental and paleonotological (particularly morphological) evidence, which is perhaps not surprising as there is no reason to suppose that the rates of evolution of morphology and molecules are equal. It is therefore worth noting the cautions expressed by some experts. Galtier *et al.* (2009) when considering the application of mtDNA analysis to the evolutionary reconstruction of species note that mtDNA is 'far from neutrally evolving and certainly not clock-like', while Ballard and Whitlock (2004) note 'it is not safe to assume *a priori* that mtDNA evolves as a strictly neutral marker because both direct and indirect selection influence mitochondrial'. Nonetheless, in the 53rd supplement to the American Ornithologists' Union check-list (Chesser *et al.* 2012) the linear sequence of orders has been changed, with Falconiformes and Psittaciformes being placed immediately preceding Passeriformes, a change 'reflecting the close relationship among these orders'.

But whatever their origins, True Falcons have proved extremely successful and have colonised all of Earth's ecozones except Antarctica, the failure to colonise that last continent is probably due to its isolation, though whether a falcon could survive Antarctica's climate is another matter: the Gyrfalcon (the world's largest True Falcon) is an Arctic dweller, but the southern winter is far harsher than that of the northern polar region, and it is debatable whether a falcon could successfully migrate across the open water that separates Australia and New Zealand from the Antarctic coast.

It is now thought that True Falcons evolved in the Miocene, probably in Africa, which is consistent with the earliest known fossil in Europe which dates from the late Miocene/early Pliocene (about 2 million years ago – Milkovský, 2002). Fossil records of Merlin, or Merlin-like falcons are known from the Pleistocene, both from the Nearctic – Brodkorb 1964, Emslie 1985, Parmelee and Oesch 1972 and Ritchie 1980 – and the Palearctic (Tyrberg 2008).

The grouping of species within the True Falcon (the *Falco*) is as contentious as the genus, with all suggested groupings having as many adherents as critics. Below, four groups which are agreed in principle by most falcon scientists are set down. The four groups accord with the current best estimate for the evolution of the falcons as a whole, which sees the Hiero falcons branch off the ancestral line first, followed by the kestrels, the hobbies and then the peregrines. There are still complications with this four group structure, not least the number of species which do not readily fit into any of the groups.

The groups are organised here with the proviso that they are presented as much as a way of listing the *Falco* as of examining the way in which the genus is truly composed. And, as we shall see, even that apparently clear-cut assemblage of just four groups has problems.

The four groups of the True Falcons comprise:

The Hierofalcons

Gyrfalcon (*Falco rusticolus*), which has a circumpolar Arctic distribution;
Lanner Falcon (*Falco biarmicus)*, of southern Europe and north Africa;
Saker Falcon (*Falco cherrug*), of central Asia;
Laggar Falcon (*Falco jugger*), of India.

The name hierofalcon was coined by Otto Kleinschmidt (1870-1954)[1], a German pastor and ornithologist who developed the theory of 'form circles' (self-contained, stable units of 'hidden nature'). Kleinschmidt probably meant 'hiero' to derive from the Greek for 'hawk' because of the usual, hawk-like, hunting strategy of the larger falcons as opposed to the stooping technique favoured by peregrines. However, in view of the pastor's creationist views it has been suggested he might have favoured the Greek for 'sacred'.

An alternative name for the group is 'desert falcons' because of the species' preferred habitat. The Gyrfalcon is a true Arctic species, so the definition of desert has to be expanded to include the polar desert of the tundra. Despite most specialists now considering that the Hierofalcons include just four species, unscrambling their evolution has not been easy (see Potapov and Sale, 2005 and references therein) and is still debated: the Prairie Falcon has also been included on occasions, although that species would appear much more closely associated with the Peregrines (Wink and Sauer-Gürth 2000, Nittinger *et al.* 2005). In another complication, Wink *et al.* (2004) consider that the Brown Falcon of Australia (see below) is associated with the Hiero group.

Nittinger *et al.* (2005) also consider that Africa was the probable origin of the Hierofalcons, while in a subsequent paper (Nittinger *et al.* 2007) suggests that the Lanner is the closest species to the African ancestor of the group, subsequent phases of emigration slitting off into Eurasia and south Asia to form the Gyr, Saker and Lagger falcons. Potapov and Sale (2005) suggest that the Gyr and Saker falcons were initially a single species, separation by the growth of the taiga forming the two species we see today.

1. For more information on Kleinschmidt and his ideas see, for instance, Potapov and Sale 2005.

The Peregrines

Peregrine Falcon (*Falco peregrinus*), of all continents apart from Antarctica;
Barbary Falcon (*Falco pelegrinoides*), of North Africa, the Middle East and west central Asia;
Prairie Falcon (*Falco mexicanus*), of North America.

Given the range of the Peregrine, the number of sub-species which have been identified across that range (20 are identified in the superb new book on the falcon – White *et al.* 2013a), the debate about the authenticity of those sub-species (White *et al.* 2013a sensibly decline to get involved with the various pros and cons of the sub-species in favour of dealing with the current 'agreed' status), and the sheer majesty of the bird, it is not surprising that copious material has appeared in the scientific press regarding the Peregrine's evolution and family tree. White *et al.* (2013a) present a brief, but cogent, assessment of the position and reference the relevant papers.

Here it is sufficient to note that some authorities consider the Barbary Falcon to be a sub-species of the Peregrine, not a full species, and that the debate over the position of the Prairie Falcon – Peregrine-Prairie hybrids are known to occur naturally – is ongoing. Very recently, further genetic analysis (White *et al.* 2013b) has suggested that an ancestral falcon may have evolved into both the Prairie Falcon and another line which itself then branched, producing one branch which was ancestral to the Hiero and Peregrine falcons, and another which was ancestral to the Kestrels, Hobbies and, presumably, other falcons which do not comfortably fit the four group pattern.

The Hobbies

Occasionally referred to as a sub-genus, Hypotriorchis. The sub-genus consists of four tree-nesting species:

Eurasian Hobby (*Falco subbuteo*), of central Eurasia;
African Hobby (*Falco cuvierii*), of central Africa;
Oriental Hobby (*Falco severus*), ranging from north-east India to Indochina;
Australian Hobby (*Falco longipennis*), of Australia.

Some experts consider that the following two, cliff-, rather than tree-nesting, species are also hobbies:

Sooty Falcon (*Falco concolor*), of the Middle East;
Eleonora's Falcon (*Falco eleonorae*), of the Mediterranean coasts of Europe and Africa.

The behavioural and morphological similarities between these two and the Eurasian Hobby are supported by molecular data (Wink *et al.* 1998) and certainly imply a close relationship. Wink and co-workers suggest that an ancestral *Falco* branch gave rise to the Merlin and a branch which then split, with the Sooty Falcon taking one path, while the second path branched again to form the Eurasian Hobby and Eleonora's Falcon.

The **Bat Falcon (*Falco rufigularis*)**, of central and South America is also very hobby-like: it has similar plumage, having red leg feathers and vent, a characteristic shared by the 'true' hobbies and by Eleonora's Falcon, though not by the Sooty Falcon, and a similar hunting technique and diet, with males specialising in hunting insects. However, it nests in tree holes rather than the stick nests favoured by the Hobbies. The **Aplomado Falcon (*Falco femoralis*)** of South and Central America and the southern USA has also been considered to be allied to the Hobbies by some. Olsen *et al.* (1989) linked the Aplomado to the Hobbies on the basis of feather protein similarities (but suggested other falcon linkages which are not favoured by more recent genetic studies). The Aplomado has red leg feathers and vent, and there are also behavioural similarities: Fiuczynski and Sömmer (2000) noted the similarities between the prey and urban breeding habits of the Aplomado (in Rio de Janeiro) and Eurasian Hobby (in Berlin).

The Kestrels

Eurasian or Common Kestrel (*Falco tinnunculus*), of Eurasia and Africa;
Lesser Kestrel (*Falco naumanni*), of southern Europe and central Asia;
American Kestrel (*Falco sparverius*), of North and South America;
Greater or White-eyed Kestrel (*Falco rupicoloides*), of southern and eastern Africa;
Fox Kestrel (*Falco alopex*), which ranges across central Africa;
Seychelles Kestrel (*Falco araea*), which is endemic to the Seychelles;
Mauritius Kestrel (*Falco punctatus*), which is endemic to Mauritius;
Madagascar Kestrel (*Falco newtoni*), of Madagascar, Mayotte, Comores, together with an Aldabran sub-species;

Moluccan or Spotted Kestrel (*Falco moluccensis*), of Indonesia;
Australian or Nankeen Kestrel (*Falco cenchroides*), of Australia and New Guinea;
Grey Kestrel (*Falco ardosiaceus*), of central Africa;
Dickinson's Kestrel (*Falco dickinsoni*), of central Africa;
Madagascan Banded Kestrel (*Falco zoniventris*), which is endemic to Madagascar.

Some experts also include the following species within the Kestrel group:

Eastern Red-footed or Amur Falcon (*Falco amurensis*), of China;
Western Red-footed Falcon (*Falco vespertinus*), of eastern Europe and west central Asia;
Red-necked or Red-headed Falcon (*Falco chicquera*), of India.

The Kestrels are a complex group, with both rufous and grey forms. Though many kestrels are to be found in Africa, which Groombridge *et al.* (2002) considered the evolutionary home of the species, the Eurasian Kestrel is the most widespread form, a falcon which breeds on Atlantic coasts in Europe and above the Arctic Circle in Fennoscandia, but is also found in Japan and South Africa. Groombridge *et al.* (2002) used genetic analysis to study not only kestrel evolution, but the interdependence of various African/near African sub-species. The work supported the idea of an African origin for kestrels with an ancient move to the Nearctic and a more recent move from Africa towards Madagascar and on to Mauritius and the Seychelles, though the timing of the movement to Australia is ambiguous.

However, while the idea that kestrels evolved in Africa fits with the distribution of species, it is not the only interpretation, complicating factors being the position of the Red-footed falcons which may be related to the American Kestrel and the fact that the Lesser Kestrel is so distinct from the other Palearctic kestrels that it seems basal to all.

The four-group classification presented above leaves several species adrift, these forming distinct continental or country groups:

Orange-breasted Falcon (*Falco deiroleucus*).
Aplomado Falcon.
Bat Falcon, of central and South America, if the Bat and Aplomado falcons are not grouped with the hobbies.
Taita Falcon (*Falco fasciinucha*), a rare species of east and southern Africa which has occasionally been placed with the Hobbies by some experts, but more recently has been suggested as being part of the

Peregrine group as a consequence of mitochondrial DNA sequencing (White *et al.* 2013b, Bell et al. 2014).

Black Falcon (*Falco subniger*), which is endemic to Australia;
Brown Falcon (*Falco berigora*), which is endemic to Australia;
Grey Falcon (*Falco hypoleucus*), of Australia, and New Guinea (?)

The Brown Falcon has occasionally been placed with the Hobbies by some experts.

New Zealand Falcon (*Falco novaeseelandiae*), which is endemic to New Zealand. The falcon is one of only two diurnal birds of prey in New Zealand (the other being the Australasian or Swamp Harrier (*Circus approximans*)) and has characteristics which suggest it fills the role of both falcon and sparrowhawk.

The above groupings leave a single falcon species, the **Merlin (*Falco columbarius*)**, which is found in both northern Eurasia and the northern reaches of North America. Merlins share characteristics with the Hobbies and the Peregrines, but are distinct from both. The position of the Merlin is complicated further by the fact that while most falcons are birds of temperate climates (warm and damp maritime, warm and dry continental) the Merlin is a cold climate species: as we shall see in *Chapter 6* male and female Merlins have adapted to the cold of their chosen habitat in different ways: males by increasing the insulation properties of their feathers, females by increasing their thermal neutral zone (i.e. the range of temperatures over which the basal rate of heat production of the bird is in equilibrium with heat losses – see *Chapter 6* and work of Warkentin and West 1990).

While the Peregrine Falcon is found as far or further north than the Merlin, the only other cold climate specialist falcon is the Gyrfalcon. The Peregrine Falcon's northward range extension appears to be less a specialism than a natural by-product of the species' astonishing success as a predator, the species breeding on all continents except Antarctica, a feature it shares with only five other terrestrial species – Barn Owl (*Tyto alba*), Cattle Egret (*Bubulcus ibis*), Glossy Ibis (*Plegadis falcinellus*), Great Egret (*Ardea alba*) and Osprey (*Pandion haliaetus*) see **Box 1** overleaf. (It should be noted though that Peregrines are largely absent from South America and equatorial Africa.) Having adapted to the cold Arctic climate the sub-species *F. p. tundrius* (the Tundra Peregrine Falcon) has exceeded the specialist Merlin in terms of northerly breeding records, perhaps in some areas even competing with the Gyrfalcon for furthest north breeding.

Genetic analysis has shed some light on the evolution of the Merlin (Wink and Sauer-Gurth 2000), constructing a family history which sees a branch of an ancestral falcon dividing to create two arms, one of which resulted in the kestrels, the other branching again, one arm forming the Hiero and Peregrine falcons, the other being ancestral to the Hobby group and the Merlin. The analysis of Wink and Sauer-Gurth also confirmed the opinion of Wink *et al.* (1998) who suggested that the Nearctic and Palearctic forms of the Merlin were now separate species rather than sub-species. Wink *et al.* (1998) used an assumed 2% sequence divergence per million years molecular clock to calculate that separation of the two forms occurred about a million years ago. DNA therefore supports the idea of speciation which some experts had already seen in the morphology, size and distribution of the two forms, though the size and morphological differences do not seem as clear-cut as some would suggest.

Box 1
Although the Pergrine and Barn Owl breed on six continents, they do not show a uniform pattern of breeding on those continents, a strange pattern for terrestrial species (though not as unusual in maritime species). The reason for such a 'staggered' breeding pattern is unknown, but is likely to be due to ecological factors or competition by established species which have prevented occupation.

Prior to the use of genetics to offer information on the evolution of species, Temple (1972b) had considered the history of the Nearctic Merlins. He considered that the available evidence suggested a Palearctic origin, with migration to the Nearctic as recently as the Pleistocene. At the time of Temple's work it was assumed there were four sub-species in North America (see *Chapter 3: Sub-species*) and his findings suggested that this number should be reduced to three. Temple also considered why the Palearctic Merlins were so different from *F.c. columbarius*, yet so similar to *F.c. richardsonii*, and proposed a history based on the periodic Quaternary glaciations of continental North America. Temple proposed an initial immigration of Palearctic Merlins during the Illinoian Glaciation when the falcons could cross Beringia as sea levels had fallen because of the extensive North American ice sheet so that Beringia temporarily linked Siberia and Alaska. During the Sangamnon interglacial which separated the Illinoian and Wisconsin glaciations, Merlins spread across northern

Opposite
Brood of five chicks of Eurasian Merlin in a ground nest in northern England.

North America. The returning ice forced the Merlins south, Temple suggesting that eastern and western populations were separated. During the Wisconsin Glaciation, Temple suggests that lobes formed in the ice, an ice-free area between these allowing a further immigration of Palearctic Merlins across Beringia which penetrated the Great Plains and, in time, became *F.c. richardsonii.*

When the ice finally retreated, the two separated populations of the original immigration moved north and linked across northern Canada, creating the range of *F.c. columbarius*, while *F.c. richardsonii* consolidated its position on the prairies. Temple's theory has much to commend it, but further work, particularly the genetic analysis of the three sub-species, will be required before it can be accepted as a true history of the Nearctic Merlin as the idea of an influx of Palearctic Merlins only a few tens of thousands of years ago does not fit well with the work of Wink *et al.* (1998) which sees the separation of New and Old World forms as occurring at least one million years ago.

Female Eurasian Merlin feeding chicks, Belarus.
Vladimir Ivanovski.

Having begun this general look at the falcon family with references to falconry, it seems appropriate to end it by returning to the topic and quoting from the Book of St Albans, a source for a recent, well-known book and film in English which makes a falcon a central character (**Box 2**). Published in 1486 the Book of St Albans was the last of only eight from the St Albans Press, which had been established in 1479 in the Benedictine Monastery of St Albans, England. The book was a compilation of information of interest to gentlemen of the day covering the topics of falconry, hunting and heraldry. The falconry section is the supposed work of Dame Juliana Barnes or Berners ('Dame' in this context meaning Prioress, rather than the feminine equivalent of 'Sir' in British social hierarchy). Though no reference to the existence of a prioress of that name at Sopwell Priory, which lies close to St Albans, exists, Sopwell's records are not complete so it is possible that Juliana existed, perhaps becoming the Priory's head as such positions were often taken by the daughters of nobility. Dame Juliana does not offer much in the way of viable information for would-be falconers, and her list is not always easy to unscramble as some of the spellings do not allow an unambiguous definition of the raptor in question, while some are clearly wrong – vultures are not the chosen falconry bird of Emperors, three Peregrines of differing types are mentioned, and most dukes would not be at all content to be given a Bustard as a sporting bird. Her list must also surely not be seen as indicating which noble had to have which bird. It was, rather, a list of avian raptor hierarchy set against the hierarchy of English nobility, a list which would likely have been crucial in the class-steeped society of the time. And so we read that a King must have a Gyrfalcon, for a Prince it must be a Peregrine, a Lady must have a Merlin, for a Young Man it must be a Hobby, while a Kestrel is for a Knave.

Box 2
The St Albans list, or more specifically the idea of a Kestrel for a Knave, became famous after the publication in 1968 of a book of that name by British author Barry Hines, and the subsequent film of the book, entitled *Kes*, directed by Ken Loach and released in 1969.

Some General Characteristics of Falcons

Falcons are diurnal raptors, adapted for predation. The eyes are enlarged to such an extent that they cannot be manoeuvred in the way human eyes can be, the falcon needing to move its head in order to change its field of view. Falcons also have two fovea (from the Latin for a pit) or positions of visual acuity, rather than the one in the human eye. One fovea is positioned to give forward binocular vision, the other positioned at an angle of about 40° to the axis of the bird which is probably associated with the tracking of prey. The function of the fovea in the hunting of the falcons is looked at when considering the flight of Merlins.

The upper mandible (maxilla) of the beak is decurved, and hooked at the tip for tearing at the flesh of prey. The tomia (cutting edges) of the maxilla have a distinct notch which creates the tomial tooth; this tooth is used to sever the spine of the prey by biting the base of the neck. This manner of killing differs from hawks (members of the Accipiter family) which tend to kill by squeezing their prey with their talons, killing by either asphyxiation or by penetrating the skin and reaching vital organs, thus causing death by multiple body trauma. The muscles which control the falcon beak are large, giving the falcons a powerful bite, allowing them to deal with prey larger than themselves. Use of the bill to kill is inherited rather than learned, appearing to be an automatic characteristic when faced with live prey. That said, adult falcons overcome this auto-function during the fledgling stage of their offspring's development as they occasionally deliver live prey for the young to deal with themselves.

Falcons also have hooked claws (talons) on their toes which are used to seize prey in an unremitting grip which may actually be fatal (causing death as in hawks), even if the main purpose is to ensure capture. The toes also have a ratchet-like tendon which means that once prey is gripped no muscular effort is required in maintaining the grip, a useful attribute when prey must be carried away to a nest or secure place for consumption.

The talons may also be used to deliver a blow to prey during a stoop: if the blow causes the talons to rake across the prey, this may cause severe damage, perhaps even death. Prey stunned by the blow, or injured by the talons usually fall to the ground and after a sharp turn the falcon will reach it and deliver a final fatal bite. The blow delivered by the feet of a stooping falcon has long been the subject of debate – is the foot closed or open? How does the bird avoid injury? The best available evidence is that the toes are extended so that the blow is associated with a potential raking by the talons. Extended toes also allow the prey to be grabbed if contact allows. But why falcons do not injure their toes or legs more frequently is not entirely clear.

Falcons share the design details of all birds – the fusion and elimination of some bones, the hollowing of others to reduce weight (but with the use of strut-reinforcing to increase strength), and a respiratory system which maximises oxygen uptake – so as to allow flight. The True Falcon wing is the classical design for rapid flight, the leading edge outboard of the wrist being swept back and tapering to a point: the falcon wing shape defines it as a hunter, rather than a soaring raptor which searches for prey on the wing.

Avian flight is complex, the mathematics used to explore it no less so, the mathematical problems being compounded by the fact that flight is more than mere wing shape and beat frequency, flying being very energetic and so involving biology as well as physics. For interested readers, the books of Pennycuick (2008) and Videler (2005) represent good starting points. Here we briefly discuss the basic features of falcon flight. In a classic paper on the flight of avian predators, Andersson and Norberg (1981) defined six flight characteristics which are important in the capture of prey: linear acceleration in flapping flight; maximum speed in horizontal flapping flight; terminal dive speed; maximum rate of climb; angular acceleration; and turning ability. The last two define the manoeuvrability of the bird. What Andersson and Norberg found was that of the six, only terminal dive speed was improved by body mass, all the others varying with the inverse of mass, i.e. smaller birds fare better (see **Box 3** overleaf). Higher body mass allows a stooping falcon to accelerate faster under gravity, the bird closing its wings to reduce its profile and so reduce drag, although this higher body mass also serves to reduce the rate of climb as the bird gains height initially. While stooping is, consequently, most usually associated with the larger falcons, Merlins also utilise it as a hunting technique.

Of particular interest in the work of Andersson and Norberg was the finding that in the two characteristics which affect the manoeuvrability of an avian predator, wing loading was important, for while falcons feeding on avian prey have high wing loading which underpins fast flight, high wing loading decreases manoeuvrability. Falcons which feed on insects therefore have lower wing loadings and can make tighter turns as the radius of the turn is proportional to the wing loading. Lower the wing loading further and lift is increased, allowing some falcons to soar when they are searching for prey, and kestrels to hover (kestrels also have longer tails in relation to other falcons which also aids hovering, as shorter tails are seen on faster fliers), a useful ability when searching for the mammals that are the primary food resource of the birds.

Wing loading also affects take-off speed as a lower number means the bird can take off with minimal effort (in comparison, say, to swans and other water birds which have to run across the water in order to generate the speed necessary to produce the lift required to get airborne). A quick take-off aids initial acceleration, though low body mass is then required for improved acceleration in flapping flight. For a surprise hunter such as the Merlin, becoming airborne quickly and without frantic wing beating which might alert the prey, and accelerating quickly is of enormous value (**Box 3**). The wing loading of male Merlins is $0.30 g/cm^2$, that of females $0.33 g/cm^2$. For comparison the figures for Peregrines are $0.52 g/cm^2$ (males) and $0.66 g/cm^2$ (females) – data from Cade (1982).

The Merlin's flight action is the most rapid and clipped of all falcons, wing-beats alternating with short glides during which the wings are held close to body. This, together with the rather square-cut tail causes distinct waves in the flight track which are further exaggerated by the obvious acceleration before glides. It was this pigeon-like flight pattern which caused the falcon to be called the Pigeon Hawk in North America. While hunting, the Merlin often climbs steeply, or may make a sudden turn after flushed prey. They may also soar, hang in wind and even hover, though such behaviour is seen infrequently.

Box 3

While the work of Andersson and Norberg (1981) gives a good general understanding of the flight characteristics of falcons and other avian predators, a note of caution must be sounded for two reasons. Firstly, falcons have to be able to stop their prey and then carry it away (while eating at the point of capture is acceptable to an adult bird, breeding requires prey to be returned to the nest), both of which require a body mass which is comparable to that of the prey. Secondly, as already noted, there is more than physics to falcon flight, and while body mass is undoubtedly important (and wing loading probably more so) the power input to flapping flight might well influence linear acceleration and horizontal flight speed positively more than body mass influences it negatively. Hence the comment that for Merlins a quick take off and fast acceleration are important in surprise hunting. Physiology may also have a greater influence over the latter than body mass alone might indicate: a heavier bird may be able to accelerate fast at the expense of earlier exhaustion, favouring short, fast attacks followed by longer resting periods, as is often seen in hunting Merlins.

Reverse Sexual Size Dimorphism (RSD)

The effect of body mass on flight characteristics is a good starting point for considering the fact that all falcons (and other raptors) show RSD, i.e. that males are smaller than females, and one theory associates this with body mass and flight ability. Indeed, the work of Andersson and Norberg (1981) was carried out in an attempt to shed light on raptor RSD.

But while raptors do exhibit RSD it is worth noting initially that despite what is occasionally seen in popular writing, RSD is not exclusive to raptors. In several other bird families – e.g. Stercorariidae (the skuas/jaegers), Scolopacinae (the sandpipers) and Sulidae (the boobies/gannets) – males are also smaller than females with no suggestion of predatory behaviour. It is also noteworthy that Laniidae (the shrikes), which predate birds, rodents and lizards, do not exhibit RSD.

Andersson and Norberg (1981) suggested that RSD in raptors had evolved as a consequence of three factors. They suggested that role partitioning favoured one sex being larger to guard the nest while the other hunted for the pair and their offspring. This does not, of itself, explain why RSD in raptors favours larger females, but once this first factor was accepted, a second factor, that it made more sense for the female to be the nest guard, did suggest RSD. The reasoning was that the female risked injury to her eggs during prey attacks, and that her role in egg laying predisposed her to be close to the nest. Once her role as a guard was established, a larger female was a more formidable opponent. The third suggestion of Andersson and Norberg was that as smaller body mass favoured better hunting (as their work on flight characteristics had shown) males of species which hunted avian prey would tend to be smaller. In favour of this Andersson and Norberg pointed out that RSD was more pronounced in species which attacked avian prey. This is true of the British falcons, where the size differential of Peregrines, Hobbies and Merlins is greater than that of the rodent-feeding Kestrel.

RSD has been the subject of a considerable number of papers over the years, both before and after the work of Andersson and Norberg (1981), but no consensus has emerged, either on the reasons for it or on whether it emerged in ancestral falcons because males became smaller or because females became larger (though data to support either would be difficult to obtain). In excess of 20 theories of RSD have been put forward, with varying levels of success in terms of both explaining the phenomenon and in achieving a degree of measured scientific acceptance. The 'most successful' theories fall into three main groups. The first is ecologically based, the idea that the sexes operate in specific niches, in this case prey size, to avoid competing with each other on a shared territory. Evidence in support of this theory is contradictory, though it should be noted that this evidence refers only to the breeding season and the theory would apply equally well during the non-breeding season. In recent years, this particular topic has been the source of a debate between researchers in India and Australia. Pande and Dahanukar (2012) noted that their work on Barn Owls in India indicated that the mean mass of prey brought to the nest by males was significantly lower than that brought by females. However, males brought a greater number of prey items. Pande and Dahanukar considered their data supported an ecological basis for RSD. Olsen (2013), replying from Australia, pointed out that a study across all raptors showed that while males did indeed carry lower mass prey than females in many species, in others the mean mass of the sexes was the same and for some, males carried heavier prey than the larger females. Replying, Pande and Dahanukar (2013) agreed that Olsen's comments were valid and called for more research into the nature of the relationship between RSD and foraging in males and females in raptors.

In Britain during the breeding season, in separate studies on Peregrine Falcons, which show a high degree of RSD, Treleaven (1977) in south-west England could find no difference in the mean size of prey delivered to the nest by the two parent birds, while Parker (1979) in Wales, and Martin (1980) in north-west England noted that the prey delivered by males was significantly smaller than the prey of females. In the most recent study Zuberogoitia *et al.* (2013) who observed Peregrines in northern Spain found no difference in the size of prey delivered to the nest by male and female falcons. In other studies it has been noted that in single chick broods the male does much of the provisioning for the youngster, while in broods of three or four the female brings more of the prey: as she is larger she is able to carry larger prey to the nest which aids feeding. This adds a further argument to the ecological theory, positing that females are larger so that they can carry larger prey to the nest: in some studies it has been noted that males consistently bring small prey items to the nest even if they are killing larger prey, for instance grouse, which they feed on. The ecological theory therefore has much to recommend it – but it does not explain why females are larger than males rather than vice versa.

The second RSD theory is behavioural, suggesting that females are larger so they can dominate males, and so maintain the pair bond and ensure food deliveries; because larger females can outcompete conspecifics in competition for males; or because males compete for females and smaller males can make superior (more agile) display flights.

The third hypothesis is physiological, the idea being that smaller males, being more agile, are better hunters and so better providers for their mates and broods, or that larger females can produce larger eggs and/or clutches and so enhance breeding success. An adjunct to this idea was the 'starvation hypothesis' which posited that larger females were better able to withstand food shortages during the breeding season. A physiological basis for RSD is certainly supported by the observation by Newton (1979) that in female falcons body condition is critical for reproduction and that females are at their heaviest during the laying phase of the breeding cycle and remain heavy (though weight is lost) during incubation and the early nestling phase. Although it is in the reproductive interests of the male to feed his mate and so increase her weight, the increased size of the female would also allow her to dominate her mate, 'bullying' him into hunting. In a study of Tengmalm's Owl (*Aegolius funereus*) Korpimäki (1986) found no evidence to support the idea that larger females are superior nest guardians, but some support for the 'starvation hypothesis', and also for the idea that smaller, more agile males are preferentially chosen by females.

More recently, further evidence for a physiological aspect to RSD has come from the work of a Norwegian group. The work began with a report by Slagsvold and Sonerud (2007) which noted that the ingestion rate of prey was variable, and that this might influence dimorphism. Ingestion rates were higher for smaller prey and for mammalian rather than avian prey. Mammalian prey was also ingested faster by raptors which were primarily mammal feeders. Slagsvold and Sonerud argue that the differing roles of breeding raptors, the male hunting, the female at the nest feeding the young, offers a potential solution to the ingestion rate problem. In later studies (Sonerud *et al.* 2013, Sonerud *et al.* 2014a, Sonerud *et al.* 2014b) the Norwegians noted that video evidence from a number of nests of Eurasian Kestrels suggested that the assumption that females delivered larger prey to the nest was flawed. It seemed that females intercepted larger prey items which they then consumed as well as feeding them to the chicks, while the male delivered smaller, more easily handled prey items directly to the chicks once they could handle these. The change from the male passing food to the female and directly provisioning the chicks occurred first for insect prey, then lizards, mammals and birds, and occurred earlier for small mammals than for larger mammals. The female's ability to intercept prey deliveries depended on her size (allowing her to bully the male) while the male's ability to return quickly to hunting also depended on his size, lending support to the contention that ingestion rates may have been influential in the development of RSD.

While it is a personal opinion, of the numerous papers to date one of the more interesting ones resulted from the work of Krüger (2005) who collected data on 237 species of Accipitridae (hawks), 61 species of Falconidae (falcons) and 212 Strigiformes (owls) and used comparative studies of 26 variables (including physiological – body mass, wing length etc. – behavioural – breeding system, displays, etc. – and ecological –

habitat, range size etc.) to investigate potential correlations with RSD. Krüger found slightly different correlations between the three raptor forms, but in all three the dominant correlation appeared to be in support of the third of the hypotheses mentioned above, i.e. that small males evolved so as to efficiently hunt small agile prey. Krüger's findings suggest that the trend for smaller males occurred before falcons began to attack larger prey, which would explain why the differential in size is more apparent in the larger falcons.

However, while the work of Krüger (2005) seems to narrow the choices in terms of RSD theories, it is likely that the debate on the topic will continue.

Food Caching

All falcons cache food, and the behaviour is not restricted to wild birds as falconry birds will also cache despite their meals being delivered regularly (assuming, of course, that the bird is aware that this will be the case). Since cached food will, if discovered, be eaten by the discoverer, falcons are furtive when caching, moving the food if, presumably, they believe they have been watched or if they believe the hiding place offers inadequate protection against discovery. Once cached, the falcon will observe the site from a short distance, presumably to fix it in its memory. Caching is used when prey is abundant, particularly if the falcon catches several prey items, for instance if taking chicks from a nest, if catching several juveniles in a flock, or if catching prey which is more abundant at certain times of day (e.g. bats as they emerge at dusk, or certain seabirds which fly far out to sea to feed). It is also used in winter allowing the bird to benefit prior to or after enduring the long hours of darkness: it may also be useful if winter weather prevents hunting. Caching may also be used occasionally as a form of food pass, a male caching prey in full view of a female, then departing.

Gastroliths

Falcons consume grit or small stones, officially known as gastroliths, but frequently termed rangle by falconers. While this is common behaviour in herbivorous birds where the gastroliths grind the forage, the reason for it in falcons is less well-understood, though the fact that it is necessary for the health of the birds has probably been known since the beginnings of falconry: Symon Latham, whose books on falconry, published in the early 17th century, are still considered to be among the best ever produced, wrote 'washed meat and stones maketh a hawk to fly'. Present opinion

Opposite
Female Eurasian Merlin bringing prey to a brood in an artifical nest in Scotland.

favours the theory that gastroliths aid the cleaning of the stomach of mucus or the grease from prey, or to aid shedding of the gizzard lining. They may also, of course, assist in grinding the food. Though well-known in falconry circles, the use of gastroliths by wild falcons is assumed, but rarely observed, though Albuquerque (1982) did note it in a female Peregrine overwintering in Brazil.

Live Prey

Any observer who has spent time with breeding Peregrine Falcons will know that the adult birds will occasionally drop not only dead, but live prey as a teaching aid for their young. Instances also occur (pers. obs. and conversations with others) of adult Peregrines 'herding' a pigeon into a quarry which holds the nest site and then circling to ensure the pigeon did not escape while the young Peregrines practise their hunting skills. In a study of the use of live prey Spofford and Amadon (1993) investigated known instances, not only of its use, but of other instances where live prey has been seen in nests, and also curious adoptive behaviour by falcons and other raptors. Spofford and Amadon (and references therein) noted instances where prey had blundered into raptor nests and also where prey had been brought in alive, but had escaped before being killed and fed to the chicks. In most instances provision of live prey is avoided because, Spofford and Amadon suggest, with elevated nests any struggle between nestlings and prey might result in the nestlings falling to their deaths. There is also no evidence that raptor chicks 'play' with prey in the manner, for instance, that domestic cats do, so there is no specific advantage in the practice. Spofford and Amadon also report instances where raptors had adopted chicks which they had brought to their nests and not killed, mentioning several instances in which Bald Eagles (*Haliaeetus leucocephalus*) had adopted Red-tailed Hawks (*Buteo jamaicensis*). In another instance a pair of American Kestrels whose brood had apparently been predated, ousted adult Starlings (*Sturnus vulgaris*) from a nest box and reared the young Starlings on a diet of mice.

Spofford and Amadon conclude that only Peregrines are known to definitely use the dropping of live prey as a teaching method. However, the behaviour has also been seen in New Zealand Falcons and is suspected (anecdotally) in other falcons.

Pellets

Though falcons usually prepare their food, plucking birds and removing the fur of mammals, some feathers and fur will be consumed and will be added to remaining indigestible parts – teeth and bone, claws and bill, the chitinous parts of insects – which cannot easily pass through the digestive tract. This detritus is formed into a pellet within the bird's gizzard, pellets being regurgitated at regular intervals, the regurgitation having the added advantage of purging the upper section of the digestive tract. (Pellets are known as casts by falconers, the action of regurgitating being called 'casting'). Pellets are normally regurgitated at dawn and as the bird is then at its roost or at the nest in the breeding season, this aids recovery of many pellets for analysis. Careful separation and analysis of the contents of the pellet reveals the prey consumed prior to its production.

Nesting

As already noted in *Chapter 1*, the True Falcons do not build nests, using the nests of other birds in which to lay their clutches, making a scrape on the ledge of a cliff or on the ground. The utilisation of these various forms varies across the Merlin's vast range, with stick nests dominating in North America, ground or ledge nesting in Iceland, and ground nests in Britain, with a combination of sites being found in Fennoscandia and Asia. More recently Merlins have taken to artificial nests, these having been used in various countries across Eurasia.

Female Icelandic Merlin.
Jóhann Óli Hilmarsson.

There is a kynde of falcon that is called a Merlyne. These Merlynes are very much like a haggart falcon in plume, in beake and talons. So as there seemeth to be no oddes or difference at all be'twixt them, save onley in the bignesse for she hath like demeanure, like plume, and very like conditions to the falcon, and in her kind is of like courage, and there must be kept as choycely and as daintily as the falcon.

Assuredly diverse of these Merlynes become passing good hawkes and verie skilfull, their propertie by nature is to kyll Thrushes, Larkes and Partridges. They flie with greater fierceness and more hotely than any other hawke of praye. They are of greater pleasure, and full of courage.

I like it well that men flie with a caste of Merlynes at once at the Larke or Lenet. For over and besides that they of themselves love companie, and to flie together, they do also give greater pleasure or delyght to the looker on. For nowe that one (at the stouping) strikes the birde, then that other ayt her downe come: and when that one clymeth to the mountie above the Larke, then that other lyeth lowe for her best advantage, which is moste delectable to beholde.

From George Turberville's *The Booke of Faulconrie or Hauking for the onely delight and pleasure of all Noblemen and Gentlemen: collected out of the best authors as well Italians as Frenchmen*, which was first published in 1575. Turberville was a scion of a Dorset family which was the likely inspiration for the d'Urbervilles in Thomas Hardy's *Tess of the d'Urbervilles*. George was a well-educated man of letters who was once an emissary to the court of Ivan the Terrible. In the extract from his book above, a *haggart* (more usually *haggard*) is the Peregrine Falcon.

The Merlin

Merlin – probably from the old French *esmerillon*, a small artillery piece, perhaps alluding to the bird's fast attack on prey. However, the root is the Germanic *esmeril*, from which the French *esmerillonné* – quick or sprightly – and *esmerveillé* – marvellous, wonderful – also derive, so the name could allude to the local view of the falcon for its ability to chase down prey. Another view is that the original French word derives from 'Stone-Falcon', used to describe the species' habitat of rocky uplands. Though this derivation has less support, the name 'Stone Falcon' was that given to the bird by islanders of the Shetlands, which lie north of the British mainland, who also noted its preference for boulder-strewn hillsides: at one time Merlins were relatively common on the Shetlands, several place names deriving from *smyril* the Norse name for the falcon.

Whatever the derivation, the French evolved into the Anglo-Norman *merlun* and the present name. The second part of the bird's scientific name *Falco columbarius*, is from 'dove', and the old American name for the species is Pigeon Hawk. The natural assumption that the bird is named for its common prey is erroneous as Merlins rarely take species as large as pigeons: the name is actually from the similarity of the Merlin to a pigeon in size and flight pattern. In Ireland the falcons were, and occasionally still are, called Snipe Hawk or Bog Hawk because of its partiality to prey and habitat.

As noted in *Chapter 1*, as a species the Merlin is curious, standing outside the defined groups of falcons (although the groups are the subject of debate, there is a majority consensus on the general grouping structure) and inhabiting a part of the world which makes it a specialist in an environment which is shared by only two other falcons, each both physically much larger and with a less specialised diet. It breeds as far north as the Arctic coast of northern Fennoscandia and western Russia, throughout Iceland, and almost to the Arctic Ocean coast in North America (Sale 2006 and Maps overleaf) though it is uncommon in all areas.

Merlins are small, but stocky, falcons with pointed wings which are relatively short in comparison to those of other small falcons, a long tail, and a small bill. When perched, the wings are clearly shorter than the tail, reaching about 75% of tail length. Merlins are sexually dimorphic, males 25-30% smaller than females. They hunt avian prey with fast, chasing flights achieved by rapid, powerful wing beats.

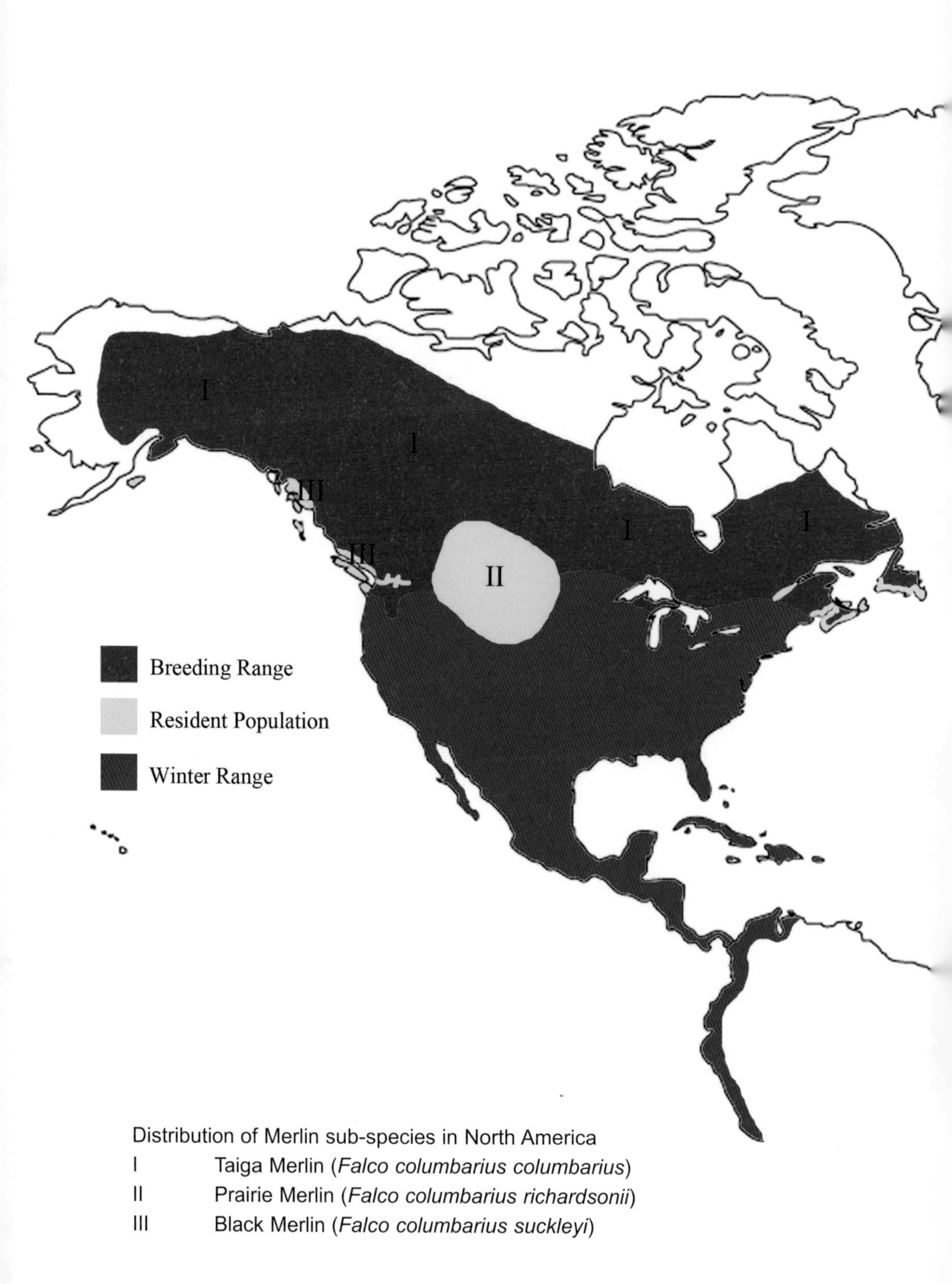

Distribution of Merlin sub-species in North America

I Taiga Merlin (*Falco columbarius columbarius*)

II Prairie Merlin (*Falco columbarius richardsonii*)

III Black Merlin (*Falco columbarius suckleyi*)

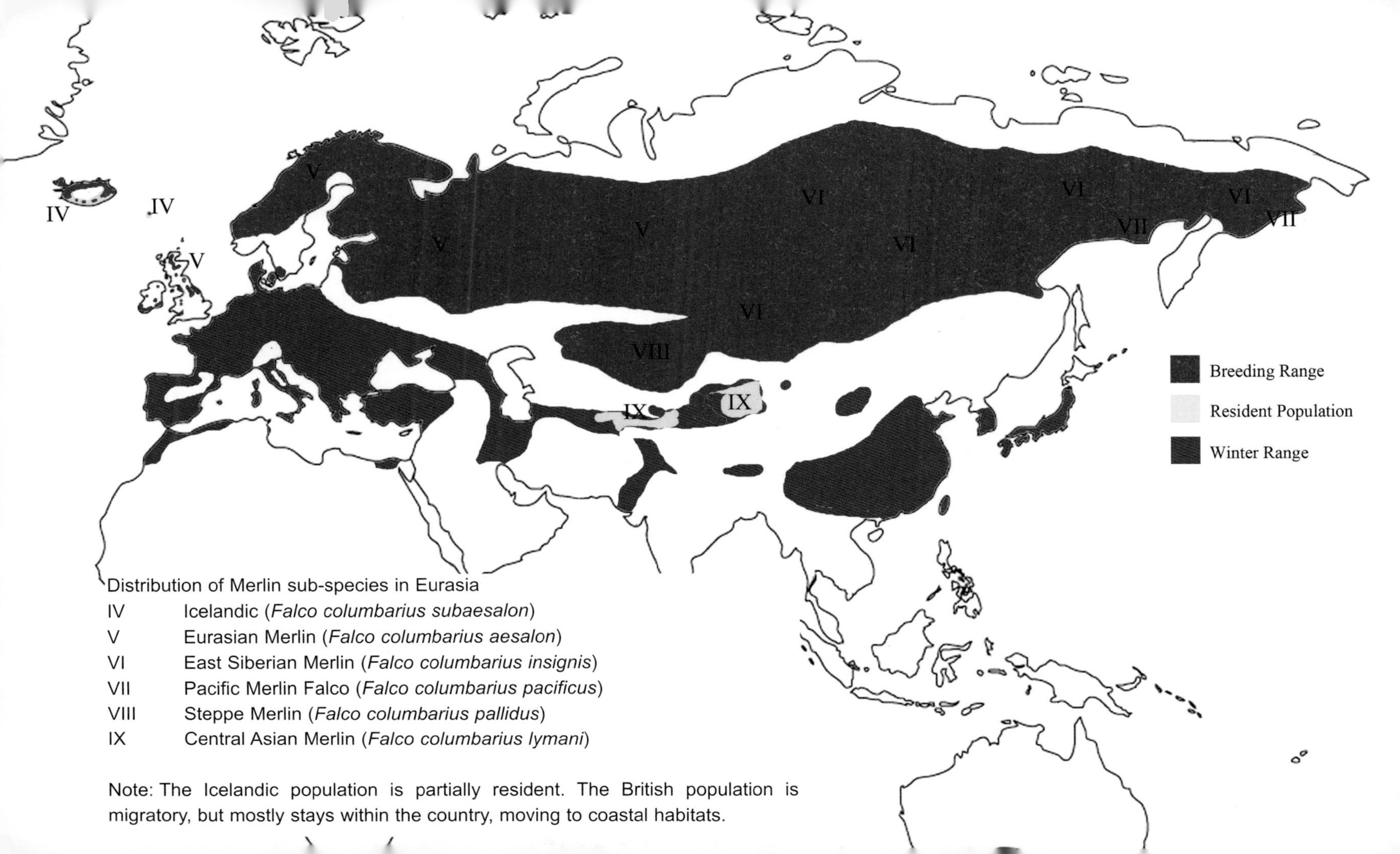

Distribution of Merlin sub-species in Eurasia

IV	Icelandic (*Falco columbarius subaesalon*)
V	Eurasian Merlin (*Falco columbarius aesalon*)
VI	East Siberian Merlin (*Falco columbarius insignis*)
VII	Pacific Merlin Falco (*Falco columbarius pacificus*)
VIII	Steppe Merlin (*Falco columbarius pallidus*)
IX	Central Asian Merlin (*Falco columbarius lymani*)

Note: The Icelandic population is partially resident. The British population is migratory, but mostly stays within the country, moving to coastal habitats.

Plumage

Merlins show a greater degree of plumage dimorphism than all other falcons apart from Kestrels. Brief details only are given below.

Nominate Adult Male

Head has blue-grey crown and auriculars, with well-defined white supercilium. Nape has paler mottling. There is no distinct head pattern as in some other falcons, but a weakly-defined malar 'moustache' is usually visible. Blue-grey back and scapulars, tending to be paler towards the rump. Throat is white, the rest of the underside pale brown or rufous-brown with dark brown barring, more prominent towards the tail. Upperwing is blue-grey, the primaries darker with occasional small, paler spots. Underwing primary and secondary feathers are blue-grey, the coverts brown. All these feathers are heavily white-spotted which gives the impression of a pale underwing with heavy dark barring when the bird passes overhead. Uppertail is dark blue-grey, almost black, with several (usually four) narrow pale bands, including a terminal band. Undertail is black with several narrow pale bands. Feathers of leg are brown or rufous-brown. Cere and orbital skin are yellow, tarsi and feet are orange-yellow.

Nominate Adult Female

Head has mid-brown crown and auriculars, with well-defined pale brown supercilium. As with male, the nape is paler mottled. Malar stripe more prominent than in male. Back and scapulars are mid-brown tending to grey-brown towards the rump. Throat is pale brown as is the remaining underside, and with prominent darker barring. Upperwing is mid-brown with darker brown primaries, the latter with small pale spotting. Underwing is mid-brown, the flight feathers darker. As with male, feathers are heavily pale spotted to give appearance of dark barring. Uppertail is brown or grey-brown with several narrow pale bands. However some birds have no bands while others have partial bands, the bands extended across only the inner retrices. The undertail is dark brown with several paler narrow bands. All undertails are fully banded, no matter what the patterning of the uppertail. Cere and orbital skin are yellow, tarsi and feet are yellow, often orange-tinged.

Nominate Juvenile Male

Head, upper body and upper wings are mid-brown, with greyish sheen, some pale spotting on the nape, and some small pale brown spotting on the flight feathers. The underbody begins with a white throat, the remaining areas being pale brown or buff with copious dark streaking which gives the appearance of heavy barring. The under wings are very dark brown, with spots of pale brown or rufous-brown which again create heavy barring. The uppertail is brown, with a darker sub-terminal band, and has two or three complete, or partial, grey bands, including a terminal band. The undertail is dark brown with several pale bands, including a terminal band. Cere and orbital skin yellow, initially tinged with green, tarsi and feet are yellow.

Although their colouring is similar to that of adult females, juvenile males differ from them in that having darker upper parts. The upper tail bands are also greyer, but this can be difficult to spot in the field. Juvenile males are, of course, also slightly smaller. The possible reason for juvenile males to closely resemble adult females is discussed in *Chapter 7: Pair Formation.*

Nominate Juvenile Female

Upper body, wing and tail are closer in colour to adult female (and therefore paler than juvenile male). Upper tail differs from juvenile male in being uniformly brown rather than having darker sub-terminal band.

After a study of 1500 museum specimens, covering all three North American sub-species of Merlin, Temple (1972a) suggested that adult and juvenile females could be distinguished by the difference in the rump and upper tail coverts, these being grey-brown in contrast to the remaining brown upperparts, while in juvenile females the rump and upper tail coverts were the same colour as the remaining back. However, Warkentin *et al.* (1992b), assessing the plumage of Merlins trapped in Saskatoon, Canada during studies of *F.c. richardsonii* (studies which will be referred to again in various sections below) found that two of nine juvenile females showed some grey in rump plumage, while 12 of 25 adult females had a brown rump. Warkentin and co-workers therefore suggest that the onset of rump coloration is not as fixed as the study of Temple (1972a) suggests.

Juveniles attain full adult plumage during their second year, though some adult females take longer to attain full adult colour on rump and upper tail coverts.

Chick

See Chick Growth in *Chapter 8*.

The differing colours of the various Merlin sub-species are considered in Distribution overleaf.

Distribution

Merlins have a circumpolar distribution, being found in all the nations which border the Arctic Ocean, and in Iceland, though absent from Greenland and the Svalbard archipelago. The plumage of sub-species tend to follow Gloger's Rule (**Box 4**).

Box 4
Constantin Gloger set down his rule in 1833, though it had already been suggested some years earlier by another German zoologist Peter Pallas. The rule suggests that darker sub-species tend to inhabit humid environments, paler sub-species being found in more arid areas. The reason is debated: it could be a result of the increased resistance of darker feathers to feather-degrading infestations, these increasing in warmer and/or wetter conditions, or for enhanced camouflage. Clarity is not aided by the numerous exceptions to the rule.

Sub-species

Studies of mitochondrial DNA have revealed genetic differentiation between various taxa, allowing evolutionary trees to be developed, as noted in *Chapter 1*. One such study (Wink *et al.* 1998) suggested divergence between the Nearctic Merlin (*Falco columbarius columbarius*) and the Palearctic Merlin (*Falco columbarius aesalon*) about 1.5 million years ago (i.e. prior to the earliest fossils so far discovered which can be confirmed as Merlin). This, together with further work (Wink and Sauer-Gürth 2000), has suggested that the continents are home to separate species as opposed to sub-species. However, this division is still being debated and here the idea that the Nearctic and Palearctic forms are sub-species has been maintained.

Gray (1958) records two instances of presumed natural hybrids between male Merlins and female Eurasian Kestrels, each dating to the 1890s. In captive (falconry) birds, hybrids are common, one of the more popular being the Peregrine x Merlin or Perlin.

In North America four sub-species of Merlin were recognised from the 1930s and formalised by the American Ornithologists Union in 1957. According to this the Merlins of boreal North America were divided into two sub-species, *F. c. columbarius*, which occupied the eastern continent, east of 95°W, and *F. c. bendirei* which occupied the rest of northern Canada and Alaska. However, this division was questioned as early as the 1930s when Swarth (1935) noted that the *bendirei* and *columbarius* were phenotypically indistinguishable. Swarth also questioned whether the forms *bendirei* and *suckleyi* were different. To shed light on the assignments, Temple (1972b) studied the skins of 1585 Merlins gathered between 15

May and 15 September covering the whole of the Merlin's Nearctic range. Temple measured the lengths of wings, tails, tarsi, toes and culmen, and calculated wing loading for the specimens. He also measured colour using a colour analyser on randomly selected skins. For each skin a reflectance curve covering the wavelength range 400-700nm was constructed. From this dominant wavelength, brightness and purity were calculated. Temple then compared the data across the geographical range of the sampled skins and concluded that while the sub-species *F.c. richardsonii* and *F.c. suckleyi* were correctly differentiated from *F.c. columbarius*, *F.c. bendirei* was not and that there was only a single sub-species covering the northern boreal zone of North America. This interpretation, with three Nearctic sub-species is now accepted.

Falco columbarius columbarius

Plumage as noted above. Often called the Taiga Merlin, the nominate breeds in Alaska, where it is uncommon everywhere. Merlins are rare in western Alaska and absent from the Aleutian Islands (though breeding on Unalaska has been claimed). Winters on Kodiak Island: there have also been claims of breeding there which suggest a small resident population. Rarely seen north of the Brooks Range and not known to breed in that area. Breeds in a wide band across Canada northward from the US border. Merlins breed in Yukon Territory and mainland Northwest Territories almost to the Arctic Ocean. Also breeds in southern mainland Nunavut. Does not breed on any of the islands of Canada's Arctic archipelago. Breeds in Quebec, but on the Ungava Peninsula, though does reach the south-eastern shore of Ungava Bay. Breeds in southern Labrador and on Newfoundland. Almost entirely migratory, but resident birds are known around the border of British Columbia and Washington/Idaho, on the southern coast of Newfoundland, coastal Nova Scotia, and along the St Lawrence River south of Quebec City. Wintering birds found as far south as Central America, islands of the Caribbean and northern South America.

F.c. richardsonii

Occasionally called the Richardson's Merlin or, more often, the Prairie Merlin.

Much paler than the nominate, though with same general pattern. Male is pale blue-grey, female pale brown. The malar stripe is absent in females, very pale and indistinct or absent in males. Male upper tail usually has three narrow dark bands in addition to the larger sub-terminal band. Juveniles are indistinguishable from females.

Found in the central 'prairie belt' of USA/Canada, and so obeys Gloger's Rule. Breeds in eastern Montana, western North Dakota, and (with smaller populations) in South Dakota and Wyoming. Also found in southern Alberta, Saskatchewan and Manitoba. In general *F.c. richardsonii* has longer wings and tail, and a lighter wing-loading than the other North American sub-species which Becker and Sieg (1987a) suggest results from the need to travel greater distances when hunting in their preferred habitat.

Note: Confusion can arise in areas where nominate and *F.c. richardsonii* overlap as nominates may also be pale. This may be due to interbreeding in these areas, though pale nominates away from the overlap region are known.

Female Taiga Merlin (*Falco columbarius columbarius*).
Nick Dunlop.

Male Prairie Merlin (*Falco columbarius richardsonii*).
Tom Grey.

F.c. suckleyi
Often called the Black Merlin.
Much darker than the nominate. Head very dark so that supercilium and malar stripe are faint or cannot be distinguished. Heavy streaking on the underside, including the throat, can give the appearance of a solid dark underside in some birds, particularly at a distance. The upper wings lack the pale spotting of the nominate and so appear solid dark blue-grey (male) or brown (female). Underwing shows pale spotting on the primaries, but secondaries are usually solidly dark, though some birds do have a sprinkling of pale spots. Uppertail has no or few (usually two) narrow paler bands and wide black sub-terminal band. Banding on undertail is clearer and may be complete or partial. Terminal band is pale.

Black Merlin (*Falco columbarius suckleyi*).
Nick Dunlop.

Opposite
Female Eurasian Merlin (*Falco columbarius aesalon*)

Males have rufous-brown tail coverts which show clearly as a contrast to the overall solid dark coloration. Juveniles are indistinguishable from females.

Confusion can arise in areas where nominate and *F.c. suckleyi* overlap as some nominate females can be very dark. However, these birds retain the pale spotting of the underwing.

Found in the humid forests of the Pacific North-West (and so satisfies Gloger's Rule). Breeds on Sitka, Admiralty and Prince of Wales islands in south-east Alaska, on the smaller islands close to them, and on the nearby mainland around Juneau. Also breeds on Vancouver Island and the nearby mainland of Canada.

Because of the small population and the relative remoteness of most population sites in comparison to those of the other Merlin sub-species, and particularly the nominate, there have been no substantive research projects on *F.c. suckleyi.*

F.c. aesalon

The Eurasian Merlin is paler than the nominate, but not as pale as the *F.c. richardsonii* . Is also smaller and less stocky than the nominate, though still a more robust bird than other small falcons. Male has more pronounced rufous-brown nape than nominate, and base colour of the underside of the male is more definitely rufous-brown. The male's uppertail has pale bands, or partial bands, or may be unbanded, but in all cases there is a broad dark sub-terminal band and white terminal band. Female upper tail usually more clearly banded than nominate, with very pale/white bands. Male birds have yellow feet tinged orange, but females and juveniles have yellow feet.

Breeds in the British Isles, Fennoscandia, Baltic states, European Russia and across Siberia to about 120°E where it overlaps with *F.c. insignis.* Reaches the Arctic Ocean coast from northern Norway through to the coast south of Novaya Zemlya, but is then confined to the inland tundra.

In Britain the Merlin breeds in the Hebridean, Orkney and Shetland islands, and throughout northern Scotland north of the Clyde, though is more scarce in the western Highlands than it is in the east. It breeds in Scotland's southern uplands and in the upland areas of northern England (Pennines, Lake District and Peak District), and in the mountains of Wales. Though once frequently seen on the moors of South-West England it is now very rare there. Also breeds in isolated places in Ireland.

In Fennoscandia, Merlins breed across much of Norway, in both upland and coastal areas, in northern Sweden, again in upland and coastal areas, and across much of Finland. In Russia *F. c. aesalon* breeds on the Kola and Kanin peninsulas, the north-east coast of the White Sea, on the Bol'shezemelskaya (Big Land) tundra to about 68°N, maintains that northern border across the Yamal Peninsula, then reaches 72°30'N on the Taimyr Peninsula. To the west the northern range boundary falls back to about 68°N as the range heads west beyond the Khatanga River. In the south Merlins are found in Latvia, northern Lithuania and northern Belarus. Further west the southern border is defined by Novosibirsk and northern Lake Baikal. For further details see Morozov *et al.* (2013).

F.c. subaesalon

Icelandic Merlins (which are also found on the Faeroes) are darker than *F.c. aesalon,* and so much closer to coloration of nominate birds. However, the underside of both males and females are white or cream, rather than pale brown and many females have a streaked throat rather than the unmarked throat of the nominate. Once considered to be differentiated from *F.c. aesalon* by being larger, particularly having longer wings, data from birds measured in northern Scotland, the Orkneys and Shetland has suggested that some Merlins in those areas were within the size range of Icelandic birds – see, for instance, Robertson (1982), who studied both the number and measured the weight and wing length of migrating Merlins on Fair Isle, an island which lies between the Shetlands and Orkneys, off Britain's northern coast. The birds found on Fair Isle were once considered to have originated from Iceland (see, for instance, Butterfield (1954) and Williamson (1954) who claimed this on the basis of taxonomy and the fact that meteorological conditions favoured an origin in Iceland or the Faeroes rather than anywhere on the 'European continent'), the possibility of them having originated on the Shetlands being overlooked. Robertson was able to show that the number of the birds recorded on Fair Isle were consistent with the number ringed on Shetland, and that their biometrics were within the ranges originally considered to differentiate the Icelandic and British sub-species. Wing length is therefore no longer considered to be a definitive measure of sub-species difference, a conclusion which has led to a re-appraisal of

Male Icelandic Merlin (*Falco columbarius subaesalon*).
Sindri Skúlason.

Icelandic bird migration data (see *Chapter 9*). However, there is an ongoing debate as to whether some Scottish Merlins (and particularly those of the Shetlands) are hybrids of the Icelandic and north European sub-species, perhaps evolving from migrating Icelandic females choosing to breed with British males rather than make the return migration. DNA sampling by Marsden (2002) indicated that there is no genetic difference between Icelandic and British Merlins, and that the observed differences may be due to geographic isolation. However, Marsden suggested caution, noting that further analysis was required before a conclusion could be reached.

Breeds on Iceland where it is the commonest Icelandic bird of prey and one of only two breeding falcons (the other being the Gyrfalcon) and the larger islands of the Faeroes, where it is the only breeding raptor. Some birds resident, though most migrate.

F.c. pallidus

Often known as the Steppe Merlin. Much paler than the Eurasian Merlin. Some males have ochre feather edging, but all have a sand/orange or pale sand/orange nape. This coloration continues on the head, though there is a blue wash to the crown and cheeks. The underside is very pale, often cream washed with buff, with only a few long, dark streaks. The underwing is pale cream with sparse dark barring and dark feather tips. Females are pale sandy brown above, with ochre streaking. The underside is pale cream with ochre streaking. Nestlings also have much paler down than the Eurasian Merlin. From the data of Morozov *et al.* (2013) Steppe Merlins are marginally larger than *aesalon*, and have a longer wing.

The range takes in the Asian steppes of Kazakhstan and western Siberia, reaching as far south-east as the Altai-Sayan Mountains. To the west the range boundary is the Ural River valley, the northern range boundary then following a line parallel to, but on the northern (Russian) side of, the Russia-Kazakhstan border east to about 75°E where there is an area of overlap with *F.c. aesalon* to the area around Novosibirsk. To the south the range border is at about the 46°E parallel. See Morozov *et al.* (2013) for further details.

Male Steppe Merlin (*Falco columbarius pallidus*).
Nigel Redman.

Male Central Asian Merlin (*Falco columbarius lymani*).
Igor Karyakin.

F.c. lymani

Very similar to *F.c. pallidus*, but males have a more distinctively buff wash to the underside, though an equally sparse darker streaking. The head is also more distinctly blue washed. Females are darker than female *pallidus*, but still paler than the Eurasian Merlin. *Lymani* Merlins are of similar size to *aesalon* but have much longer wings.

Occasionally called the Central Asian Merlin, this sub-species has a discontinuous range across south-eastern Kazakhstan and into Kirghizia where the falcons reach the Tien Shan Mountains and cross the border into China. Further east another population breeds on each side of the Altai Mountains, being more populous on the northern (Russian) side than on the southern (Mongolian) flanks. Karyakin and Nikolenko (2009) consider that 82.5% of the sub-species population probably lies within the Russian republics of Tuva and Altai, and the Altai-Sayan Mountains. Karyakin and Nikolenko consider the isolation of the Altai-Sayan birds from the remaining range of *F.c. lymani* is such that they can be considered a separate sub-species – the Altai Merlin. This suggestion awaits ratification. For further details see Morozov *et al.* (2013).

F.c. insignis

Sometimes referred to as the East Siberian Merlin. The male's head is tinged ochre, with a white forehead. The underparts are tinged pink, and the tail has poorly distinguished or no banding apart from a terminal band. The upperparts of the female are paler brown, but with a distinct red-brown patterning on the back. The underside is also spotted and streaked red-brown, this most obvious on the crural/tarsus feathering. The female tail has much more pronounced banding than the male. Juveniles are as the female, but usually paler and with less pronounced red-brown patterning or streaking. Morozov *et al.* (2013) consider the East Siberian Merlin to be larger than *F.c. aesalon*: this was based on measurements on a small sample, but each of these was significantly larger than an average *aesalon*.

Breeds, as the name implies, in eastern Siberia, ranging from the Lena Delta (with a zone of overlap westward to about 120°E) where it is found to about 72°N, eastwards across the Yana and Indigirka rivers (at about 68°N) to the Kolyma Delta. The eastern range border is assumed to be about 160°E on the Chukotka Peninsula where there is an overlap with *F.c. pacificus*. The southern range border is the northern tip of Lake Baikal

Female East Siberian Merlin (*Falco columbarius insignis*).
Andrey Kovalenko.

Female Pacific Merlin (*Falco columbarius pacificus*).
Igor Dorogoy.

and from there along latitude 56°N to the coast of the Sea of Okhotsk at Udskaya Bay, but that is conjectural, particularly as *F.c. pacificus* is also thought to have a range on the same coast to about 56°N and it is assumed there is an area of overlap. For further details see Morozov *et al.* (2013).

F.c. pacificus

The Pacific Merlin is the darkest of the Asian sub-species. Very similar to *F.c. aesalon*, though in general darker, and also larger. The underparts of both males and females are patterned with large spots.

Breeds in the extreme east of Siberia, along the Bering Sea coast from the southern Chukotka Peninsula to northern Kamchatka and along the western coast of the Sea of Okhotsk as far south as Udskaya Bay. There are reports of Merlins breeding on the Shantar Islands and northern Sakhalin, but these are unproven. However, there are rare incidents of Merlins breeding on the Kuril Islands, which link Kamchatka's southern tip to Japan, with the exact sub-species involved (*F.c. insignis* or *F.c. pacificus*) still debated (Nechaev, 1969). It was probably a *pacificus* Merlin which was spotted by the great Russian ornithologist Leonid Portenko (1896-1972) on Wrangel Island in 1939. A handful of Merlins have been seen on the island since then (Stishov *et al.* 1991). (While it might be assumed these observations and other falcons seen on the Taimyr Peninsula at about 72°30'N are the most northerly latitudes at which Merlins have been seen in fact vagrant birds have been seen on Svalbard and Novaya Zemlya – Morozov *et al.* 2013.) For further details on the exact range see Morozov *et al.* (2013).

Moult

Merlins are no different from other birds in spending a great deal of time preening their feathers, applying a secretion from the preen gland to maintain waterproofing, and ensuring that the feathers lie correctly maximising flying efficiency and thermal insulation. Wright (2005) records fledgling Merlins bathing in a stream – behaviour which was perhaps associated with ridding themselves of parasites, as Wright notes young chicks often carry Hippoboscidae flies. From other reports, bathing would seem to be rare behaviour, though Merlins have been seen to take deliberate showers in rain, extending wings and tail. However, in a study in south Wales, Haffield (2012) records that the breeding Merlins he observed bathed almost every day. Rebecca (1987) observed dust bathing, noting a female Merlin in Scotland which flew down to a patch of sand and gravel beside a busy road. The bird fluffed her feathers, pushed her underside into the 'dust', then walked slowly forward and flapped her wings, presumably ensuring the 'dust' was widely scattered. The bird dusted for a period of four minutes and was then disturbed by a car, but returned for a further two minutes of dusting before moving to a perch to preen. Similar behaviour has also been seen in North American Merlins: Sodhi (2005) noted one bird dusting seven times in an hour. The bird was parasite-infested, which may explain this particular intense session of dusting. Haak and Buchanan (2012) also report a Black Merlin bathing. This bird was also drinking, behaviour which has been observed in captive Merlin, but rarely in the wild, though Morozov *et al.* (2013) note that in the warmer climate of the steppes, Merlins often seek out water to drink. In addition to water and dust bathing, Merlins have also been observed sun-bathing, both in summer (Rolle 1999, Dickson 2003) and winter (Dickson 1998b).

However, despite the fastidious care birds take in ensuring their feathers stay in good condition, feathers are damaged, raptor feathers probably sustaining a higher level of damage than those of non-raptors because they may add collisions, struggles etc. to the normal wear-and-tear experienced as a result of landing and take-off. In addition, feathers are subject to the attention of a myriad of parasitic feather mites which degrade both the vanes and quills (as well as infesting the skin and subcutaneous tissues, blood etc.): in a review of the parasitic mites of falcons and owls Philips (2000) found a total of 21 families infesting falcons (and 17 infesting owls), including at least four species which infested the feathers, or were found on the skin of Merlins.

To maintain good feather condition, all the feathers of adult birds are replaced once annually, the requirement to maintain flight ability at all times in order to be able to hunt meaning that replacement is over an extended period.

In general the moult begins in June for both male and female, females usually beginning earlier, and is completed in the autumn. The moult pattern of the primaries is not fixed, and may start 4-5-6, 5-4-6 or 6-5-4, but is usually completed as 3-7-8-2-9-1-10. However, Morozov *et al.* (2013) state that for Steppe Merlin the sequence is 6-7-5 (or 7-6-5) 8-4-3-9-2-10-1, but provide no data for the sequence in the other three Asian sub-species.

The tail feathers are moulted from the centre outwards. The timing of the moult is critical for the birds, particularly the males, as adult Merlins have to replace their feathers while maintaining flight efficiency as they have to hunt, not only to feed themselves but their offspring. Since the fledglings of prey species are available at the time when Merlin young require significant food intake, adult Merlins exploit this time not only to feed their young but to maximise their own energy intake to fuel the moult. As noted above, females start the moult earlier, at a time when the male is hunting to feed his mate and, following hatching, the young nestlings. In a study of moulting in urban-breeding Merlins in Saskatoon, Canada, Espie *et al.* (1996) noted that the timing of the male moult depends on the quality of his territory: if prey resources are plentiful the male may start to moult earlier. The start of the male moult was also found to be correlated with hatch date.

Adult Merlins may also suspend the moult if conditions require. Espie and co-workers found this was significantly more likely in males than females, and that 48% of males (of 52 sampled birds) had suspended, an indication of the energetic overload of hunting in the breeding season, but also of the flexibility in moult which allows the male to divert energy away from moulting to feeding his offspring if that is what is required. For female Merlins Espie *et al.* (1996) found that those birds which did not arrest the moult had significantly smaller broods than those that did.

In general, Juveniles partially moult body feathers and tail feathers (occasionally only the central pair) in early spring, with other feathers at the same time as the adults so that full adult plumage is then acquired. In a study of the number of first-year birds moulting during the winter/early spring period Pyle (2005) found that only 4% of the birds were moulting in September-November, and that these were replacing only 5% of their feathers. In December-February, a third of birds were moulting, replacing 15% of their feathers. By March-May all birds were moulting, with 50% replacement.

Dimensions

Overall Length (bill tip to tail tip) of all North American sub-species male 272.0±6.8mm; female 302.6±8.1mm; Wing: male 199.4±4.0mm; female 221.9±4.0mm. Tail: male 121.6±2.6mm; female 136±3.4mm.

F.c. subaesalon is bigger, with male wing 208.0±4.7mm, female wing 227.0±2.6mm; male tail 122.0±3.4mm, female tail 134±2.6mm.

F.c. aesalon is comparable to the North American sub-species, with male bill 12.4±0.4mm, female bill 13.8±0.6mm; male claw 10.8±0.3mm, female claw 11.8±0.4. See, also, further details in Table 1.

Attribute	Male		Female	
	Mean	Range (Sample Size)	Mean	Range (Sample Size)
Wing length (mm)	199	193-205 (12)	222	206-231 (45)
Tail length (mm)	120	117-124 (9)	133	120-142 (42)
Culmen (mm)	13.7	12.5-13.8 (8)	15.0	14.1-17.0 (38)
Tarsus width (mm)	3.1	3.0-3.3 (10)	3.6	3.2-3.9 (44)
Tarsus depth (mm)	4.2	3.7-4.7 (10)	4.9	4.4-5.5 (44)
Tarsus length (mm)	36.5	33.5-38.0 (10)	38.5	36.0-38.5 (44)
Toe (mm)	28.2	25.0-30.5 (10)	32.5	29.3-31.9 (44)
Weight (g)	174	163-188 (13)	247	227-277 (46)

Table 1 Measurements of adult *F.c. aesalon* Merlins trapped in northern England. From Wright (2005).

In his study of Merlin's on Britain's Orkney Islands, Picozzi (1983) measured some of the physical data noted in Table 1 for fledglings no older than five days from fledge and noted that on average the male tarsus length, width and depth were 35.1mm, 3.4mm and 5.0mm respectively, with females measuring 38.0mm, 4.2mm and 5.9mm. The mean mid-toe was longer for males and females at 34.3mm and 37.4mm respectively. Picozzi also measured the hind claw, his sample of newly fledged birds having a mean length of 10.4mm (male) and 11.4mm (female).

From the data of Morozov *et al.* (2013) the Asian Merlin sub-species have dimensions as follows:

F.c. pallidus
Male wing 208.4±0.8mm
Female wing 227.8±1.4mm

F.c. lymani
Male wing 229.7±1.8mm
Female wing 248.3±2.3mm

F.c. insignis
Male wing 204.0±0.9mm; tail 130.6±1.5mm
Female wing 224.5±1.0mm, tail 139.4±1.8mm

F.c. pacificus
Male wing 209.1±0.9mm; tail 135.3±1.8mm
Female wing 221.2±1.5mm, tail 147.9±2.0mm

As can be seen, the Central Asian Merlin (*F.c. lymani*) has much the longest wings of the sub-species.

Weight

North American male 166.9±11.1g, female 255.5±17.1g.
F.c. aesalon is comparable, *F.c. subaesalon* on average 30g heavier.
Data on the Asian sub-species is from a sample too small for meaningful comparison.

In all cases the weight is variable throughout the year and particularly during the breeding season. The changes in weight, and the reasons for these changes, are discussed when considering Breeding in later Chapters.

Habitat

The Merlin's breeding habitat varies across its vast range. In North America nominate birds prefer open, or largely open, areas, with boreal birds breeding close to forest clearings, frequently near rivers, lakes or bogs and occasionally on islands in larger lakes. In upland areas the birds choose areas which replicate these boreal conditions, breeding in forested areas, but hunting over the extensive open areas between. On the North American prairies Merlins are not found in wheat-growing areas, as these do not attract sufficient prey species. In such areas Merlins are absent

unless there are areas of open grassland. In a study across the range of *F.c. richardsonii* (Prairie Merlins) Fox (1964) found that mixed woodland, river edges and islands were the preferred habitat, with 84% of utilised stick nests being in deciduous trees (aspen, elder, willow and birch), the remainder in conifers (chiefly White Spruce (*Picea glauca*) and Jack Pine (*Pinus banksiana*)). However, in a study of *F.c. richardsonii* in south-eastern Montana, Becker and Sieg (1987a) noted that large areas of open grassland were not a preferred habitat. Becker and Sieg identified five habitats which were utilised by the falcons: sagebrush/grassland, riparian, stands of Ponderosa Pine (*Pinus ponderosa*), grasslands and agricultural areas. But statistically, the first three habitats were used more frequently than random chance would suggest, the latter two less frequently, implying that open grassland was not, of itself, a preferred option. Further study suggested that in all the potential habitats, areas of patchy scrub/open grass were the preferred hunting areas – and these were provided more extensively in the first three identified habitats.

As noted above, the preferred habitat of *F.c. suckleyi* is the humid forests of the Pacific North-West, though again areas of open land are required for hunting.

In Iceland the birds breed on the coastal plain avoiding much of the central island which is heavily glaciated and where vegetation is sparse and so potential prey is uncommon. The coastal plain holds a mix of undeveloped lava fields and agricultural land, each of which is a habitat for potential prey species. Iceland as a whole is largely treeless, though some small areas of forest have been planted and maintained over recent years.

In Britain, Merlins are confined to remote undeveloped areas of rough grazing or moorland, with some agricultural areas. Rebecca (2011) identified the habitat types across Britain based on the principal habitat found within 1km of the falcon's nest area. Rebecca found that dry heather moor was most common in Britain as a whole, with mixed dry and wet moorland the next most common. Only south-west Scotland and Wales, where Merlins were more likely to be found in coniferous woodland, were exceptions, though even there dry moorland was the 'second favourite' habitat.

During migration the birds are usually found on farmland, particularly arable land. Newton *et al.* (1978) studied Merlins nesting in Northumberland, northern England, an area which includes the large Kielder Forest, which then measured 30km by 20km. No nests were discovered deeper than 1km from the forest edge, emphasising the

Opposite
Typical Merlin habitat at the treeline in northern Canada.

Above Muskeg on Admiralty Island, Alaska, typical habitat for Black Merlins.
Below Merlin habitat in north-west Iceland.

importance of open land to the species for hunting. The birds also ignored clear-felled areas if these were deep within the forest.

In Fennoscandia and much of European Russia, Merlins breed in areas which combine the aspects of North American and Icelandic/British habitats, i.e. a semi-open habitat, moorland or heath, with scattered trees which allow the possibility of utilising corvid stick nests: ground nesting is rare in this area, though it does occur if the habitat is good, but trees are sparse or absent. River valleys and bogs are preferred areas because these are favoured by the passerines which form the main diet. Fennoscandian Merlins do not nest in dense forests, particularly avoiding the commercial spruce plantations which dot upland areas, but will nest at forest edges if there are available stick nests and the local open areas offer the opportunity to hunt. As in Britain, most occupied areas tend to be rough, undeveloped land.

In the Baltic nations and European Russia the preferred Merlin habitat is as in Scandinavia. Further east in Siberia the three northern sub-species (*F.c. aesalon*, *F.c. insignis* and *F.c. pacificus*) are found in similar habitats, preferring shrub-tundra, both lowland and at higher elevations, areas of forest-tundra and sparse mountain forests. The latter include the thin birch forest of the Polar Urals, though not at heights above about 300m. Merlins are only absent from extreme northern lands where the bare tundra of the polar desert offers insufficient resources for prey species to thrive, though Merlins are found on the vast northern deltas of the great Siberian rivers where wader populations can be hunted. In more southerly latitudes the three northern Merlins breed in the shrub-tundra and sparse wooded areas in river valleys. One interesting development, mirroring the move to urban dwelling seen in North America, *F.c. aesalon* has been seen not within, but on the outskirts of several cities – Riga and St Petersburg to name the most well-known, but several less well-known cities as well (Morozov *et al.* 2013 and references therein).

The Steppe Merlin (*F.c. pallidus*) is a falcon of the steppe and forest-steppe, particularly river valleys, but breeds on land converted to agriculture, nesting in tree belts planted as windbreaks to protect pasture and agricultural land. In their study in the Altai-Sayan region of Russia, where three sub-species *F.c. aesalon*, *F.c. lymani* and *F.c. pallidus* breed, Karyakin and Nikolenko (2009) noted that *F.c. aesalon* were found in sparse larch forests at the edge of the steppe, while *F.c. lymani* preferred sparse, high Siberian Larch (*Larix sibirica*) forests along river valleys. In such places (which represented over 85% of identified habitat) the Merlins were found at altitudes up to 2613m, the highest recorded for any sub-species: the mean height of nests of these 'Altai Merlins' was 2017±307m (range 872m-2613m). At lower altitudes the falcons were found in larch forests close to the steppe, but such 'lowland' Merlins were

Moorland Merlin habitat in northern England.

a minority (3.8% of the total): the remaining Merlins were found in high willow forest or high mixed willow/larch areas. *F.c. pallidus* were found in dry steppe areas with significant areas of sub-saline lakes, and scarce trees, usually as pine forests with birch tree borders. In all cases the nests of the Merlins were found no more than 50m from the forest edge.

In Yorkshire, northern England, studies over long periods where records have been analysed for nest sites used over a period of at least 20 years (Rowan 1921-22, Part I), reveal that even if the Merlins at a particular nest site have been killed, the site is very likely to be re-occupied (data supported by Newton *et al.* 1978). Rowan's study was carried out at a time when the killing of raptors on land reserved for game bird shooting was common and indiscriminate and he gives two remarkable examples: in one case both birds of a pair of Merlins were trapped and killed every year from 1898 to 1916 at the same nest site in northern England, despite the fact that no adult bird escaped killing and no egg was hatched; in his other case Rowan reports a similar story over a 12-year period. These examples indicate the importance of both habitat and nest site to the species, but, as Rowan notes, does raise the question of where the continuous supply of new Merlin pairs came from. In a separate study of Merlins in northern England, Newton *et al.* (1978) noted that of 37 areas known to have been used in the period 1880-1940 at least 25 were still in use in their study period 1971-1976. Of the 12 that were not used, five were no longer suitable for Merlins as a consequence of change in land usage or increased human presence. This site fidelity is not restricted to British Merlins, Trimble (1975) noting that a ground nest site in Newfoundland had been used continuously for 23 years. However, it is important to distinguish between the fidelity of Merlin as a species to specific requirements in terms of nest site and territory and the fidelity of individual birds. The latter is discussed in *Chapter 8: Breeding Territory and Mate Fidelity*.

While the records of Rowan (1921-22, Part I) and Newton *et al.* (1978) indicate the importance of habitat to Merlins, in recent years the habitat preferences of the species have evolved to include urban areas, at least in North America. James (1988) notes that prior to 1971 such urban dwelling/breeding was unknown (or, at least, unrecorded). In that year a pair of Merlins bred in Saskatoon: by 1987 the population had increased to 27 pairs. James puts the appearance and increase in Merlins down to the increase in American Crows (*Corvus brachyrnychos*) and Black-billed Magpies (*Pica hudsonia*). Being omnivorous and opportunistic, the corvids had invaded urban North America once trees planted there had matured sufficiently to provide good sites for stick nests, the Merlins taking advantage of the nests in later years (**Box 5**). Within urban areas Merlins are able to take advantage of flocks of House Sparrows (*Passer domesticus*) as prey, the fact the passerines are available year-round allowing some Merlins to overwinter in towns and cities despite temperatures regularly reaching -35°C. In winter the Merlins roost in conifer stands which give some protection against night-time temperatures. In a study of the trees used Warkentin and James (1990) used data from radio-tagged birds and chance observation to assess preferred roosts of Merlins in Saskatoon. Warkentin and James used a series of 14 parameters to define trees – species, height of roost, diameter of tree at breast height, and others which defined the number and height of other trees local to the roost tree – in an attempt to define preferred roosts. They found no difference in the requirements of males and females. In each case the Merlins chose

Box 5
Oliphant and Haug (1985) noted that in addition to the availability of both stick nests and prey in urban areas, Saskatoon had the advantage of the South Saskatchewan River which runs through the city, Richardson's Merlins being invariably found on the major river systems of the prairie belt. It is probable that only similar rivers would give rise to other urban Merlin populations, which might explain the absence, to date, of urban-dwelling pairs in Eurasia.

conifers, White Spruce or Blue Spruce (*Picea pungens*), which were taller and had greater crown vegetation than random trees, and were close to other trees. The Merlins tended to roost on the leeward side of the tree. The preferences indicate the need for the falcons to gain shelter from prevailing winds, and so reduce convective heat loss – being on the leeward side, and nearby trees acting as a windbreak, will help reduce this loss, as will greater crown cover which will also reduce radiative heat loss. The height of the roost is more likely governed by a desire to reduce predation. However, given the apparent desire to gain shelter from the wind, it is interesting that the study did not show any inclination for the birds to use buildings as shelter, either directly by roosting on or within them, or by choosing trees in the lee of them. Perhaps the relatively short time that Merlins have been city dwellers has not allowed them to recognise a potential benefit of urban living.

Staying with urban-dwelling Merlins, James (1988) noted that the move to the new habitat had resulted in not only a change of diet for the falcons

An area of the Altai-Sayan region of south-central Russia, close to the borders of Kazakhstan and Mongolia, where *F.c. aesalon*, *F.c. pallidus* and *F.c. lymani* may all be seen.
Igor Karyakin.

but a change of behaviour: rural Merlins are noisy, giving their alarm call at the sight of a human, but in urban areas they are silent. As James notes the Merlins 'would do little else if they reacted to every human being that passed by their nest trees.' James also notes that the Merlin invasion of urban North America has been encouraged in some cities due to concerns over the decline in rural populations following the problems caused by organochlorine contamination. Within the urban environment the Merlins hunted in gardens, parks and cemeteries (Sodhi and Oliphant 1992).

As interesting as the fact that Merlins now breed in urban environments in North America are the further results of the study of Sodhi and Oliphant (1992) which showed that movement to the city is gradual, at least as regards the hunting range of the birds (see *Chapter 7: Territory*).

Voice

Merlins are very vocal during the breeding season, but much quieter at other times.

The alarm call, usually written '*kee-kee-kee*', is made by both male and female, the male voice higher pitched and more rapid. However, the present author does not consider the call to be harsh enough to be written with a 'k', considering a softer first sound, '*hwee-hwee-hwee*', though in what follows the more general construction is followed. For observers in North America, the call will be reminiscent of that of the American Kestrel, though delivery is much faster, while in Europe the same comment can be made with respect to the Eurasian Kestrel. Craighead and Craighead (1940) state that if the birds were highly excited their call ended with a piercing '*ki-ki-ki-ki-kieeee*' or guttural '*kac kac kac*' (and state that 'once heard it can hardly be mistaken'). Occasionally a bird will intersperse single '*kee*' calls between full alarm calls. Feldsine and Oliphant (1985) identified the common alarm call, but also three others. A '*chip*' call is made by both males and females during courtship, and is often heard by females if the male is not in sight. The male *chip* is again higher pitched and more rapid. A '*chrrr*' is made by males seeking copulation with females, both in pairs and by non-paired males who are intruding on a territory. Feldsine and Oliphant delightfully suggest that this sounds very similar to a short blast on a police whistle, which may require non-Canadian readers to consult a sound archive to see if they agree. Finally there is a whining call used by females begging for food from their mate. The same call may be made by hungry nestlings, but is then more urgent. Nestlings will make the same sound if they see an aerial predator, probably in an effort to alert an adult. Females will also make a sharp '*tick tick*' when they are feeding nestlings, presumably to reassure the youngsters, but seemingly as an indication of contentment.

Diet

Merlin are essentially a predator of aerial species, taking birds (typically weighing less than 40g, but as we shall see not restricted to smaller prey), but also flying insects and bats. In a study in central and western Europe Duquet and Nadal (2012) list a total of 11 bat species predated by raptors, particularly in late summer and autumn, and while the majority of these were taken by Eurasian Kestrel, Eurasian Sparrowhawk, Hobby, and Peregrine Falcon, Merlins were also found to have captured the mammals. Merlins will also take terrestrial mammals, though these usually form a small fraction of their diet. During a huge increase in the vole population of the Scottish borders in the early years of the 1890s a witness to the Commission set up to investigate the causes and effects of the 'vole plague', as it was termed, claimed that Merlins did not prey on voles, and by implication any terrestrial mammal, but this was contrary to the evidence provided by Laidlaw (1893) which showed clearly that they did. (It is probable that the rodent responsible for the plague was the Field Vole (*Microtus agrestis*) though at the time, 'voles' and 'mice' were commonly interchanged in reports so there is no absolute certainty.) Rudebeck (1951) states that from his observations Merlins were able to subsist on a diet composed mainly of lemmings when they were abundant, though noting that this was exceptional and that avian species were invariably the main prey, while Wiklund (2001) in a study in two Arctic-fringe National Parks in Sweden also provides evidence that the diet of Merlins includes rodents (voles and Norwegian Lemming (*Lemmus lemmus*)) even though the primary prey remains avian species. Indeed, Wiklund's study showed that the breeding population of Merlins increased, and Merlins bred in areas in which they were not normally found, in years when the rodent population was high. However, when during the winter of 2001/2002 the population of Far-Eastern Vole (*Microtus fortis*) increased dramatically in the Primorye region of far-eastern Russia, the number of wintering Merlins did not increase, while those of other rodent eaters did. Volkovskaya and Kurdyukov (2003) recorded an increase in Common Buzzard (*Buteo buteo*), Eurasian Kestrel (by a factor of 3), Hen (Northern) Harrier (*Circus cyaneus*) by a factor of 5 and Rough-legged Buzzard (*Buteo lagopus*), indeed the total number of buzzards rose by a factor of 11, largely due to the increase in Rough-leggeds.. It is, of course, possible that there was no local Merlin population to move into the area as most had migrated south or that the number of rodent specialists meant that the Merlins were outcompeted.

While Wiklund's data and that from the Scottish vole plague make it clear that Merlins will take rodent prey, it is equally clear from Iceland, where there are few rodents (all of which have been introduced by human settlers and are almost exclusively found in or close to human habitation) that Merlins can exist entirely (or almost entirely) on a diet of avian prey. Warkentin (1985) also found evidence of a young Merlin, recently released into urban Saskatoon, feeding on Meadow Voles (*Microtus pennsylvanicus*). Warkentin also notes other references to mammalian prey (other voles (Microtus spp), gophers (Geomyidae spp), the Thirteen-lined Ground Squirrel (*Spermophilus tridecemlineatus*) and Least Chipmunk (*Tamias minimus*), as well as several bats). It is probable that young Merlins are responsible for taking most terrestrial mammal prey as in the early days of fending for themselves they are likely to find these, and insects, much easier to catch than avian prey. Becker (1985) in a study of the diet of *F.c. richardsonii* in south-east Montana also found the remains of a Northern Short-horned Lizard (*Phrynosoma douglassi brevirostre*). Page and Whiteacre (1975) also record an unidentified lizard in the prey of a Merlin wintering on the coast of California. In his description of the Merlin, Palmer (1988) also states that unusual prey items include garter snakes, 'horned toads' (Phrynosoma horned lizards), toads, crayfish, scorpions and spiders, but gives no specifics. In a study of prey remains in north-east Scotland, Rebecca (2006) found a single Eurasian Common Frog (*Rana temporaria*).

Equally unusual, to judge by the absence of reports in the literature, is the taking of carrion. Haug (1985) reported seeing a Merlin taking road kill. In radio-tracking a Coyote (*Canis latrans*) in north-central Wisconsin, Haug came across a remarkable, if appalling, scene, with a flock of 20-30 Purple Martins (*Progne subis*) lying dead or injured on the road, having apparently been hit by traffic as they took advantage of radiating heat from the road surface. While watching, Haug saw a Merlin fly to the scene and take a martin away, returning after five minutes or so to take another, and then taking another after another five minutes. Haug surmised the Merlin was caching the martins as it was April and so too soon for there to be nestlings in need of feeding. The Merlin seemed to take injured rather than dead birds, but the report does indicate that the falcons will perhaps take fresh carrion opportunistically. McIntyre *et al.* (2009) also reports Merlins scavenging road kill, seeing two juvenile birds mantling and consuming a dead Snowshoe Hare (*Lepus americanus*) on one occasion and a female Merlin at a dead Snowshoe Hare on a second occasion. Each time the hare had been killed by vehicles on the road into the Denali National Park. There is also a report from Scotland (Thomas 1992) of a Merlin feeding on a fresh road-kill Rabbit (*Oryctolagus cuniculus*).

Prairie Merlin with a Mourning Dove.
Greg Lasley.

Wright (2005) also noted that Merlin nestlings which died, through sickness or injury, were invariably eaten, though Wright does not specify whether this was by siblings or parents. However, Dickson (1998a) definitely records young Merlin consuming a dead sibling.

In general, Merlins consume their prey on the ground, or in trees, but Kerlinger *et al.* (1983) observed one falcon consuming a small bird on the wing during a migration flight (see *Chapter 9: Migration Flights*) while Bent (1938) also notes having seen a Merlin eating a rodent on the wing.

Because of the range of Merlins it is difficult to compile a comprehensive list of prey species, as this changes from area to area, and also across the year as seasonal movements of both the Merlin and other species bring the two together, and as the number of fledglings increases as the breeding season reaches its conclusion. Consequently the number of species identified in the Merlin diet can be extensive and include surprises. Diets identified from several points across the species' range are tabulated below, with some general patterns being given in the following paragraphs.

In North America, the main prey species of Merlins vary across the range: in Alaska the prey are American Tree Sparrow (*Spizella arborea*), Dark-eyed Junco (*Junco hyemalis*), Lapland Longspur (Lapland Bunting – *Calcarius lapponicus*) and northern sparrows; at Churchill, on the southern shore of Hudson Bay, Knapton and Sanderson (1985) recorded the main prey as Savannah Sparrow (*Passerculus sandwichensis*), but also found one adult Lesser Yellowlegs (*Tringa flavipes*) as well as nestlings which may perhaps have been from that species; in rural southern Canada the main prey is Horned Lark (Shore Lark – *Eremophila alpestris*); while in urban areas House Sparrows form the major part of the diet. North American birds, especially juveniles finding food for themselves after becoming independent, also take flying insects, particularly dragonflies.

In Iceland Merlins feed on Meadow Pipits (*Anthus pratensis*), Eurasian Golden Plover (*Pluvialis apricaria*), Common Snipe (*Gallinago gallinago*) and Dunlin (*Calidris alpina*). In Britain Meadow Pipits are also taken, as are Northern Wheatears (*Oenanthe oenanthe*), Starlings and Skylarks (*Alauda arvensis*), in addition to finches and waders. British records also include Goldcrest (*Regulus regulus)*, the smallest European bird, weighing around 5g, to the Eurasian Green Woodpecker (*Picus verdis*) which is both large (up to 225g) and formidably armed. The taking of such large prey is an exception, though Merlins will take adults of larger bird species opportunistically, as well as the nestlings of larger birds. In Finland, Bergman (1961) notes that a Merlin took all 12 chicks of a brood of Grey Partridge (*Perdix perdix*). British records also include a number of day-flying moth species (particularly the Emperor (*Saturnia pavonia*) and the Northern Eggar (*Lasiocampus quercus callunae*)), dragonflies and beetles, and, but rarely, rodents. In Scandinavia Meadow Pipits again form a large proportion of their diet, together with other pipits and finches.

Several observers have reported finding the remains of the nestlings of prey species in the nests of Merlins (e.g. Sperber and Sperber (1963) who noted nestlings of Brambling (*Fringilla montifringilla*) in a study in Norway and McElheron (2005) who found that the feathers of unfledged nestlings outnumbered those of adult birds by 2 to 1 in late summer at the plucking post in his study in Ireland). However, see Table 9 which is based on data from Northumberland, England, where Newton *et al.* (1984) found that even when the percentage of fledglings in the diet was at its highest, fledglings still constituted only about a quarter of the Merlin diet. Kermott (1981) records direct evidence of nestling predation rather than reliance on prey remains. In Montana he observed a Merlin being chased by American Robins (*Turdus migratorius*). Having noted two nestlings in the Robins' nest, Kermott later noticed the Merlin returning, then being attacked again and flying away with something in its talons. On inspecting the nest again, Kermott noticed that there was now only one nestling, confirming nest predation by the Merlin.

Intuition would suggest that the composition of the Merlin diet would reflect the relative abundance of prey species, assuming that all available species were of more or less equal merit as prey, and more or less equally easy to catch. But the assumptions involved in equating abundance and

Box 6
Baker and Bibby (1987) were seeking to shed light on whether the evolution o plumage coloration was based primarily on predation avoidance or on s selection, the merits of these two theories still being debated, in part because difficulty of mounting experiments to test the theories, particularly as coloration might reduce predation by one predator could actually increase likely capture ra a different predator. The work of Baker and Bibby (on domestic cats as w Merlins) suggested that cats primarily took prey which was cryptically colo whereas Merlins seemed primarily to hunt according to prey abundance. The w Sodhi and Oliphant (1993) considered a further theory, that of optimal-foraging, suggests that prey should not be taken according to relative abundance, but re profitability, usually defined by net energy intake, i.e. that some prey provides energy input, relative to the energy requirement of capture. The overall conclus the two studies would seem to suggest that prey selection is very far from the s equation intuition would suggest, and that more work is required before d conclusions can be reached on prey selection, in either Merlins or other preda

likelihood of capture mean that intuition is not a good judge of the subtleties of prey selection. In a study in Wales, involving analysis of 6366 avian prey remains recovered from Merlin plucking sites, Baker and Bibby (1987) found that Merlin prey reflected the local abundance of prey species, but that the falcons were very slightly more likely to catch prey which was conspicuously coloured (**Box 6**). In a study of urban Merlins in Canada Sodhi and Oliphant (1993), noted that the falcons were less likely to capture cryptically coloured House Wrens (*Troglodytes aedon*), but noted that House Sparrows and Shore (Horned) Larks were taken more frequently than expected, implying that the Merlins were selecting their prey rather than merely responding to abundance. In their study Sodhi and Oliphant noted that the two preferred species were more likely to leave cover than other species, and were therefore more vulnerable, and that their mass was in the range 21-40g which was the preferred mass of the Merlins (based on the observed prey mass spectrum) (Fig. 1).

One interesting aspect of the work of Sodhi and Oliphant (1993) was that adult Merlins were observed to preferentially select juvenile House Sparrows over adults during the period when their own brood was fledging. During the incubation phase of the Merlin breeding cycle, adult male and female House Sparrows are taken as would be expected. This was also the case during the Merlin nestling phase, although juvenile House Sparrows are added to the diet, at that stage. But during the Merlin fledgling stage, more juvenile sparrows are taken. Sodhi and Oliphant note that this increase corresponds to a period when the juvenile sparrows spend more time out of cover, presumably as they seek food for

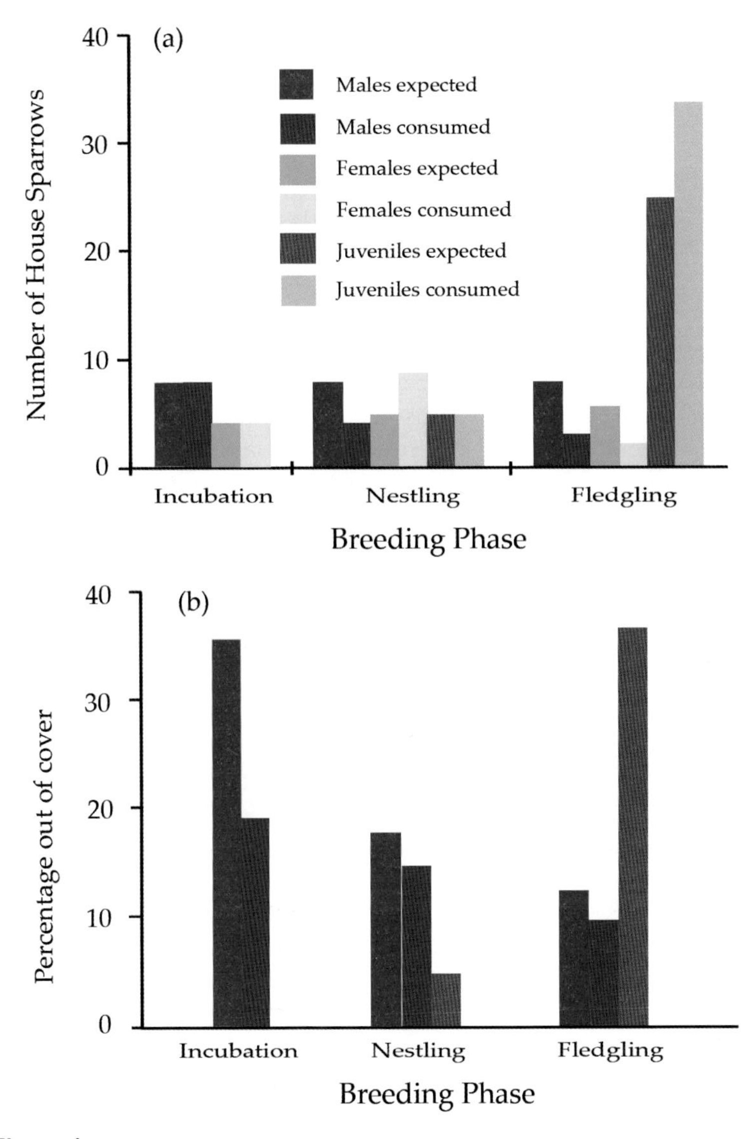

Figure 1
a) Adult Merlin consumption of House Sparrows at various phases of the breeding cycle.
b) Percentage of time spent out of cover by adult and juvenile House Sparrows during phases of the Merlin breeding cycle.
Redrawn from Sodhi and Oliphant (1993).

themselves having become independent of their parents, and so are more vulnerable to attack.

Also interesting was the study of Bibby (1987) in Wales which noted that the diversity of the Merlin diet increased as the proportion of farmland within 2-4km of the Merlin nest site increased (see *Chapter 8*, Fig. 18). Prey diversity was also reflected in the egg size of the falcons (see *Chapter 8: Eggs*).

Composition of the Diet

The composition of the Merlin diet can be ascertained in two ways. The simplest is to observe the feathers or skeletal remains (sternum, pectoral girdle, synacrum, legs) at plucking posts or the remains at nest sites in order to identify prey species and the number of each. But this method has several drawbacks, as Newton *et al.* (1984) note. Firstly, the feathers of large, pale birds are more easily seen than those of small, dark birds. Secondly, the adult Merlins may consume larger prey away from the nest as these are difficult to carry, so remains at the nest may not correctly represent the diet. The adults may also consume smaller prey, such as moths, themselves so these, too, may be under-represented in nest remains. Thirdly, the birds may entirely consume some small prey items which would therefore leave no remains.

An alternative to collecting remains at plucking posts or nests is to use pellet analysis. It is generally assumed that pellet analysis is a more accurate method of defining diet, particularly if small mammals are a significant fraction of the total prey, but in a study of the diet of Irish Merlins, Fernández-Bellon and Lusby (2011a) found there was no significant difference in the two methods for Merlins, probably because they almost exclusively consume avian prey. A combination of the two methods is, of course, also possible. In general, in the data below, the diet has been established by observation of prey remains at plucking posts or at nests, though some data does derive from pellet analysis.

Merlin pellets collected during summer by Johnson and Coble (1967) on the Isle Royal National Park in Lake Superior, and therefore for the nominate Merlin, had a mean size of 25mm x 12mm. For *F.c. aesalon*, Dickson (1999) collected both summer and winter pellets in south-west Scotland. The summer pellets had a mean length of 29.3±8.8mm (range 12.5mm-53mm, N=76)) and mean width (at the widest point) of 11.6±2.1mm (range 8.0mm-16.5mm). In winter the pellets were smaller, mean length 23.6±6.7mm (range 12-41mm, N=88), mean width 11.3±1.9mm (range 6-16mm).

In the Tables below the dietary differences of Merlins across their range are explored.

In a study in Alaska's Denali National Park, Laing (1985) recovered prey remains close to Merlin nests (Table 2). Although at least 20 bird species were identified, five species comprised almost 50% of the diet. The heaviest of the prey species is the Varied Thrush which may weigh up to 80g as adults. The thrushes collected were assumed to be adults, explaining the relatively high biomass fraction despite the low numeric fraction of prey items.

Prey Species	Percentage of total prey items[3]	Percentage of total biomass
American Tree Sparrow	11.5	8.5
Dark-eyed Junco	10.6	8.1
Lapland Longspur (Lapland Bunting)	9.8	10.7
Fox Sparrow (*Passerella iliaca*)	8.9	13.3
White-crowned Sparrow (*Zonotrichia leucophrys*)	8.0	7.9
Water Pipit[1]	6.2	5.4
Snow Bunting (*Plectrophenax nivalis*)	5.3	8.3
Savannah Sparrow	3.6	2.6
Rosy Finch[2]	2.8	3.1
Varied Thrush (*Ixoreus naevius*)	2.8	8.6
Orange-crowned Warbler (*Vermivora celata*)	2.8	1.0
Horned Lark (Shore Lark)	1.8	2.7
Semipalmated Plover (*Charadrius semipalmatus*)	1.8	3.4
Yellow Warbler (*Dendroica petechia*)	1.8	0.7
Wilson's Warbler (*Wilsonia pusilla*)	1.8	0.6
Golden-crowned Sparrow (*Zonotrichia atricapilla*)	1.8	2.4
Lagopus spp.	0.9	-
Least Sandpiper (*Calidris minutilla*)	0.9	0.9
Tree Swallow (*Tachycineta bicolor*)	0.9	0.7
Northern Waterthrush (*Seiurus noveboracensis*)	0.9	0.6
Common Redpoll (*Carduelis flammea*)	0.9	0.5
Zonotrichia spp.	0.9	1.0
Catharus thrushes	0.8	9.0
Unidentified species	5.3	-

Table 2 Diet of Taiga Merlins in the Denali National Park, Alaska. From Laing (1985).

Notes:

1. Laing quotes Water Pipit (*Anthus spinoletta*) but it is now considered that the American Pipit (*Anthus rubescens*) is a distinct species from the European species.
2. Laing gives 'Rosy Finch' the scientific name *Leucosticte arctoa*, but it is now considered that the North American species (the Grey-crowned Rosy-finch – *Leucosticte tephrocotis*) differs from the Asian Rosy-finch (*Leucosticte arctoa*) of eastern Asia.

3. Laing totals the two columns to 100%, but clearly the left-hand column does not sum to 100. There is nothing in the text to explain this discrepancy, which may be due to the left-hand column ignoring non-avian prey as a fraction of total prey, while the right-hand column assumes non-avian prey has minimal biomass.

Becker (1985) studied the prey of Merlins by examining pellets recovered in south-eastern Montana which comprised 67% grassland, 27% forest and 6% badlands, the whole area dominated by hills and buttes to 800m (Table 3). The majority of the prey species weigh 10-40g, though Western Meadowlarks and Killdeer can weigh up to 100g.

Prey Species[2]	Percentage of total prey items	Percentage of total biomass
Horned Lark (Shore Lark)	27.2	21.2
Lark Bunting (*Calamospiza melanocorys*)	17.6	21.0
Vesper Sparrow (*Pooecetes gramineus*)	12.6	9.5
Mountain Bluebird (*Sialia currocoides*)	6.6	4.5
Western Meadowlark (*Sturnella neglecta*)	4.2	12.0
Chestnut-collared Longspur (*Calcarius ornatus*)	3.7	1.9
Red Crossbill (*Loxia curvirostra*)	3.0	2.9
American Goldfinch (*Carduelis tristis*)	2.1	0.9
Chipping Sparrow (*Spizella passerine*)	1.6	0.6
Killdeer (*Charadrius vociferus*)	1.6	5.0
Unidentified birds	1.4	1.6
Bobolink (*Dolichonyx oryzivorus*)	1.2	1.1
Brown-headed Cowbird (*Molothrus ater*)	1.2	1.5
Brewer's Sparrow (*Spizella breweri*)	1.2	0.4
Brewer's Blackbird (*Euphagus cyanocephalus*)	1.2	2.4
Lark Sparrow (*Chondestes grammacus*)	0.9	0.6
Townsend's Solitaire (*Myadestes townsendi*)	0.9	0.9
Other Birds[1]	3.7	5.0

Table 3 Diet of Prairie Merlins in Montana[3]. From Becker (1985).

Notes:

1. Other birds included House Sparrow, Common Poorwill (*Phalaenoptilus nuttallii*), Red-winged Blackbird (*Agelaius phoeniceus*), American Robin, Cliff Swallow (*Petrochelidon pyrrhonota*), Eastern Kingbird (*Tyrannus tyrannus*), Savannah Sparrow, Dark-eyed Junco, Baltimore (or Northern) Oriole (*Icterus galbula*), Mourning Dove (*Zenaida macroura*) and Yellow-rumped Warbler (*Dendroica coronata*).
2. In a study of the sub-species in southern Alberta, Hodson (1978) noted that Horned Lark comprised 50% of prey, with Chestnut-collared Longspur accounting for 37%, Western Meadowlark 3%, Vesper Sparrow 2%, the remaining 8% being unidentified passerines.
3. In addition to the avian prey Becker (1985) also noted grasshoppers (4% of total prey), moths (1.2%), mammals and, as noted earlier, a Northern Short-horned Lizard (mammals + lizard = 2.8%).

Icelandic Merlin with Eurasian Golden Plover.
Jóhann Óli Hilmarsson.

In a study of urban-dwelling Richardson's Merlin, Sodhi and Oliphant (1993) collected prey remains found close to nest sites to derive the constituents of the Merlin diet (Table 4). The largest contributor by a very significant margin was the House Sparrow, adults of which weigh about 30g. While most prey species weighed less than about 40g, American Robins can weigh up 80g, and both Brewer's Blackbird and Bohemian Waxwings (*Bombycilla garrulus*) can weigh 55-65g.

Prey Species[2]	Percentage of total prey items	Percentage of total biomass
House Sparrow	64.5	55.5
Horned Lark	9.1	8.8
American Robin	4.3	10.5
Swainson's Thrush (*Catharus ustulatus*)	3.7	3.6
Cedar Waxwing (*Bombycilla cedrorum*)	2.6	2.6
Chipping Sparrow	2.2	0.9
Yellow Warbler	1.9	0.6
Snow Bunting	1.4	1.9
Brewer's Blackbird	1.4	2.8
Bohemian Waxwing	1.3	2.4
Clay-colored Sparrow (*Spizella pallida*)	0.9	0.2
White-throated Sparrow	0.7	0.5
Hermit Thrush (*Catharus guttatus*)	0.6	0.6
Others[1]	4.9	8.9

Table 4 Diet of urban Prairie Merlins in Saskatoon derived from a study 1987-1990. From Sodhi and Oliphant (1993).

Notes:

1. Other birds, which constituted less than 0.5% of the total included, in order of numbers taken, Dark-eyed Junco, Black-capped Chickadee (*Poecile atricapilla*), Least Flycatcher (*Empidonax minimus*), juvenile Rock Dove (*Columba livia*), Lapland Longspur (Lapland Bunting), Mountain Bluebird, Song Sparrow (*Melospizia melodia*), Western Kingbird (*Tyrannus verticalis*), Hairy Woodpecker (*Picoides villosus*), Barn Swallow (*Hirundo rustica*), House Wren, Killdeer, Yellow-rumped Warbler, Blue Jay (*Cyanocitta cristata*), White-crowned Sparrow, (European) Starling, Yellow-bellied Sapsucker (*Sphyrapicus varius*), Red-necked Phalarope (*Phalaropus*

lobatus), Pine Siskin (*Carduelis pinus*) and Western Meadowlark. In addition, the Merlins took Little Brown Bat (*Myotis lucifugus*) and Richardson's Ground Squirrel (*Spermophilus richardsonii*).

2. In an earlier study (Oliphant and McTaggart 1977), also carried out in Saskatoon soon after Merlins had begun to breed in the city, the falcons had also taken Ruby-crowned Kinglet (*Regulus calendula*), Red-eyed Vireo (*Vireo olivaceus*), Chestnut-collared Longspur, Tree Swallow and Lincoln's Sparrow (*Melospiza lincolnii*). The Merlins also pursued, but did not catch Northern (or Common) Flickers (*Colaptes auratus*) though Oliphant and McTaggart note that there are records of the remains of Flickers being found in Merlin nests. In a later study (Sodhi *et al.* 1991a) the Merlins also took Common Grackle (*Quiscalus quiscula*), Tree Swallow and Eastern Kingbird.

In a study by James and Smith (1987) of urban Merlins in Edmonton and Fort Saskatchewan the prey species also included Evening Grosbeak (*Coccothraustes vespertinus*), Pine Grosbeak (*Pinicola enucleator*), Red-winged Blackbird, Eastern (or Rufous-sided) Towhee (*Pipilo erythrophthalmus*), Say's Phoebe (*Sayornis saya*), White-winged Crossbill (*Loxia leucoptera*), and, very unfortunately for the little bird doubtless both enjoying, but daunted by, its probably all-too-brief freedom, a Budgerigar (*Melopsittacus undulatus*).

Prey species of North American Merlins:
Left American Tree Sparrow. *Right* Shore (Horned) Lark.

In a study of the Icelandic sub-species of Merlin, Nielsen (1986) collected remains from plucking post sites (Table 5) in a study area in the north-east of the island. The majority of prey items were from birds weighing 15-50g, but weights ranged as high as 200g (juvenile Barrow's Goldeneye).

Prey Species[1, 11]	Percentage of total prey items	Percentage of total biomass
Meadow Pipit[2]	23.4	8.0
Snow Bunting[3]	8.8	5.4
Redwing (*Turdus iliacus*)[4]	5.8	6.3
White Wagtail (*Motacilla alba*)[5]	3.0	1.0
Northern Wheatear[6]	2.2	1.2
Unidentified passerines	0.2	0.1
Eurasian Golden Plover [7]	16.2	23.4
Dunlin[8]	14.2	13.9
Common Snipe[9]	13.1	27.4
Red-necked Phalarope	5.0	3.6
Juvenile Redshank (*Tringa totanus*)	2.0	2.0
Ringed Plover (*Charadrius hiaticula*)	1.9	2.3
Juvenile Whimbrel (*Numenius phaeopus*)	0.2	0.4
Unidentified shore birds	1.3	1.3
Juvenile Barrow's Goldeneye (*Bucephela islandica*)	0.2	0.2
Unidentified ducks	0.4	0.4
Juvenile Rock Ptarmigan (*Lagopus mutus*)	0.2	0.7
Arctic Tern (*Sterna paradisaea*)[10]	1.2	1.9
Unidentified birds	0.5	0.3

Table 5 Diet of *F.c. subaesalon* in Iceland derived from a study 1981-1985. From Nielsen (1986).

Notes:

1. The avian diet, as listed in the Table, comprised 99.8% of the Merlin diet. The remaining 0.2% derived from the Wood Mouse (*Apodemus sylvaticus*).
2. The Meadow Pipit remains comprised 6.3% adults, 5.3% juveniles and 11.8% unclear.

3. The Snow Bunting remains comprised 2.8% adults, 2.5% juveniles and 3.5% unclear.
4. The Redwing remains comprised 1.3% adults, 2.0% juveniles and 2.5% unclear.
5. The White Wagtail remains comprised 0.2% adults, 0.8% juveniles and 2.0% unclear.
6. The Northern Wheatear remains comprised 0.5% adults, 1.5% juveniles and 0.2% unclear.
7. The Eurasian Golden Plover remains comprised 1.3% adults, 12.6% juveniles and 2.3% unclear.
8. The Dunlin remains comprised 11.3% adults, 1.0% juveniles and 1.9% unclear.
9. The Common Snipe remains comprised 6.8% adults and 6.3% juveniles.
10. The Arctic Tern remains comprised 0.2% adults and 1.0% juveniles.
11. In an earlier study close to Mývatn, a lake in north-central Iceland famous for its birdlife, Bengtson (1975) observed the diet of Merlins (remains collected from nest sites) as very similar to that of Nielsen in terms of both species and fractions, though with more passerines and fewer waders, as would be expected for an inland, rather than coastal, area. Mývatn is also well-known for its duck population, and Bengtson noted juveniles of Mallard (*Anas platyrhynchos*), Eurasian Teal (*Anas crecca*), Eurasian Wigeon (*Anas penelope*), Greater Scaup (*Aythya marila*), Tufted Duck (*Aythya fuligula*) and Long-tailed Duck (*Clangula hyemalis*) in the diet. He also found a small fraction of juvenile Black-headed Gull (*Larus ridibundus*).

Prey species of Icelandic and British Merlins:
Left Meadow Pipit. *Right* Northern Wheatear.

In a study of Merlins in Northumberland, north-east England, Newton *et al.* (1984) derived the prey species data of Table 6. The data were derived from over 1800 prey items recovered from close to nests or at plucking sites. Although most prey items weighed 10-40g, some were 100g.

Prey Species[1]	Percentage of total prey items	Percentage of total biomass
Meadow Pipit[2]	55.8	36.8
Skylark	11.8	15.6
Chaffinch (*Fringilla coelebs*)	5.8	4.6
Starling	3.6	8.9
Northern Wheatear	2.0	2.0
Fieldfare (*Turdus pilaris*)	1.8	6.5
Song Thrush (*Turdus philomelos*)	1.5	3.2
Willow Warbler (*Phylloscopus trochilus*)	1.2	0.4
Goldcrest	1.1	0.2
Common Snipe	1.1	3.8
Whinchat (*Saxicola rubetra*)	1.0	0.6
Common Redpoll	0.6	0.2
Ring Ouzel (*Turdus torquatus*)	0.6	2.0
(European) Robin (*Erithacus rubecula*)	0.6	0.3
Pied Wagtail (*Motacilla alba yarrelli*)	0.5	0.4
Redshank	0.4	1.7
Blackbird (*Turdus merula*)	0.4	1.3
Mistle Thrush (*Turdus viscivorus*)	0.3	0.9
Eurasian Golden Plover	0.3	2.0
Lapwing (*Vanellus vanellus*)[3]	0.2	0.7
(Juvenile) Red Grouse (*Lagopus lagopus scoticus*)	0.2	1.3
Feral Pigeon (*Columba livia var*)	0.1	0.7
Stock Dove (*Columba oenas*)	0.1	0.5

Table 6 Diet of *F.c. aesalon* in Northumberland, north-east England defined by land usage within 1km of nest sites. From Newton *et al.* (1978).

Notes:

1. As well as the tabulated bird species, the following species were found in prey remains, though in no case did these account for either 0.5% by number or by biomass: Blue Tit (*Parus caerulus),* Brambling, Bullfinch (*Pyrrhula pyrrhula*), Coal Tit (*Parus ater*), Common Crossbill (*Loxia curvirostra*), Cuckoo (*Cuculus canorus*), juvenile Curlew (*Numenius arquata*), Dunlin, Dunnock (*Prunella modularis*), juvenile Eurasian Kestrel, European Goldfinch (*Carduelis carduelis*), Greenfinch (*Carduelis chloris*), juvenile Grey Partridge, House Martin (*Delichon urbica*), House Sparrow, Linnet (*Carduelis cannabina*), Long-tailed Tit (*Aegithalos caudatus*), Redwing, Reed Bunting (*Emberiza schoeniclus*), Sand Martin (*Riparia riparia*), Siskin (*Carduelis spinus*), Snow Bunting, Barn Swallow, Treecreeper (*Certhia familiaris*),Tree Pipit (*Anthus trivialis*), Willow Tit (*Parus montanus*) and Wren (*Troglodytes troglodytes*).

Other prey items included Common Shrew (*Sorex araneus*), beetles, caterpillars, Fox Moth (*Macrothylacia rubi*) and Peacock Butterfly (*Inachis io*). An earlier study of Merlin diet, in what is now England's Lake District National Park, Brown (1935) also noted the remains of Long-tailed Field-Mouse (now more commonly called the Wood Mouse) and 'Short-tailed Field-Mouse'. It is not clear what species Brown meant by the latter, but it is likely he meant the Short-tailed Vole (now more commonly called the Field Vole).

2. In a study of Merlins in south-east Yorkshire, England, during the period 1983-2002, which collected feathers from plucking posts and remains at nest sites, Wright (2005) found that Meadow Pipits comprised 50.9% of the Merlin diet, with a 29.7% contribution from Starlings. The next most popular prey species was the Northern Wheatear, with 4% of the diet. Wright found that 17 species comprised the total diet.

3. A study of prey species in the Orkney Islands (Meek 1988) made by examining remains at Merlin plucking posts showed that one Merlin had an enthusiasm for newly-hatched Lapwing chicks, eight pairs of legs being found.

Weir (2013) examined the diet of Merlins in Speyside, central Scotland during the period 1964-1984 from the remains of 656 prey items found in three differing habitats (open moor, birch moor and pine moor) (Table 7 overleaf).

Prey Species[1]	Percentage of total prey items[2]
Meadow Pipit	80.8
Northern Wheatear	6.6
Skylark	3.4
Chaffinch	3.1
Pied Wagtail	1.4
Red Grouse	1.2
Dunlin	0.8
Other Birds[3]	2.0
Mammals[4]	0.9

Table 7 Diet of Merlins in Speyside, north-east Scotland.

Notes:
1. In addition to vertebrate prey Weir collected 687 invertebrate remains of various moths, beetles and dragonflies. The fractions of prey taken by the Merlins in the three habitats were approximately the same.
2. Weir does not differentiate between adult, juvenile and nestling prey items, but notes that in most species all three forms were taken: biomass data could not therefore be calculated by the present author.
3. Other birds included Bluethroat (*Luscinia svecica*), Bullfinch, Coal Tit, Common Sandpiper (*Tringa hypoleucos*), Fieldfare, Golden Plover, Mistle Thrush, European Oystercatcher (*Haematopus ostralegus*), Reed Bunting, Ring Ouzel, Sand Martin, Siskin, Snipe, Barn Swallow, Tree Pipit, Whinchat, Willow Warbler and Wren.

Rebecca *et al.* (1987) also record two Bluethroats among the prey remains of a Merlin in north-east Scotland. As these are rare arrivals from Scandinavia – Rebecca and co-workers note that only around four are seen each year – for local birdwatchers this was an unfortunate choice by the Merlins, though Bluethroats are regularly taken in Scandinavia (see below).

Rebecca (2004) also noted that tree-nesting Merlins in north-east Scotland took a much larger fraction of Barn Swallows than suggested by the data above, finding the remains of 51 (35 adults and 16 juveniles) in 118 prey items, suggesting that for the particular Merlin or Merlin pair, swallows represented 43% of the total diet.
4. The mammals included voles and a Pygmy Shrew (*Sorex minutus*).

Such a high predation rate of Meadow Pipits may have an impact on local populations. In a study of the impact of raptors on prey species' populations in southern Scotland Amar *et al.* (2008) noted that Meadow Pipits had declined in number by 60% over a seven-year period during which the Hen (Northern) Harrier breeding pairs had doubled from 10 to 20 within the study area: there were also 10 breeding pairs of Merlin. About 60% of the diet of both raptors comprised Meadow Pipits, Amar and co-workers estimating that one falcon pair would consume 91 pipits during the breeding season, so that in their study area the 10 falcon pairs would take over 900 pipits, about 18% of the population.

Dickson (1995), observing six nests in Scotland, noted that 0.26 prey items/hour were delivered (by the male to the female, or to the chicks) during the pre-laying phase of breeding (seven in the morning, four in the afternoon and one in the evening), this falling to 0.15/hour during incubation (the times also altering to two in the morning, four in the afternoon and five in the evening), then rising to 0.54/hour (7 morning, 18 afternoon, 12 evening). During the post-fledging phase, 0.49 items/hour were delivered to the chicks (7 morning, 11 afternoon, 21 evening). The total number of deliveries was therefore 99, consistent with the total suggested by Amar *et al.* (2008).

In Eire, Fernández-Bellon and Lusby (2011a), studied the diet of Merlins at ten sites in the counties of Donegal, Galway and Wicklow by a combination of collecting remains at plucking posts and analysing pellets. They found that 69.5% of the prey was avian, with 33.1% being moths and 1.0% dragonflies. Of the bird species, Meadow Pipit and Skylark were the main prey, though the percentages in the diet were much lower than those in Table 6, with 22.9% and 8.9% respectively (17.0% and 13.2% by biomass respectively). Irish Merlins also took more Barn Swallows (4.7% by occurrence, 3.9% by biomass) and many more Snipe (3.9% - 17.4%) and Mistle Thrush (2.9% - 14.9%). The moth fraction in the diet was very high by occurrence, but actually contributed only 3.2% by biomass. The most frequently taken species were Fox Moth, Emperor Moth and Northern Eggar, as was the case in Britain. In a study of Merlins that breed on the lower slopes of the Wicklow Mountains (to the south of Dublin, Eire, a range which reaches 925m – 3035ft), McElheron (2005) found that in spring the birds supplemented their diet of local species with incoming Barn Swallows arriving exhausted after their long flight from Africa, while in summer, the local Chaffinches were supplemented by pre-fledge nestlings taken from local nests.

Newton *et al.* (1984) looked at the variation of prey species through the Merlin breeding season. Dividing the prey into woodland and open-country species, Newton and co-workers noted that the former dominated the Merlin diet early in the season, the latter at later stages of the breeding cycle (Table 8). Newton and co-workers considered this was due to the availability of open-country prey early in the season, and this was confirmed by Petty *et al.* (1995) who recorded a similar trend in the Merlin diet from woodland to open-country species, noting that Meadow Pipits, the main prey of Merlins in northern England, were absent from the Merlin breeding habitat in winter and did not return to the moorland, on which they bred, until the Merlins had returned and established territories.

	April[3]	**May**	**June**	**July**
	N (%)	**N (%)**	**N (%)**	**N (%)**
Open country species	270 (76.9)	548 (81.4)	537 (90.7)	358 (93.0)
Woodland species[1,2]	81 (23.1)	125 (18.6)	55 (9.3)	27 (7.0)

Table 8 Percentages of open country and woodland species in the diet during the breeding season. From Newton *et al.* (1984).

Notes:
1. Woodland species include scrub and woodland edge.
2. In their study of Merlins breeding on moorland in north-east Wales, Roberts and Jones (1999) noted that while the percentage of moorland birds increased as Newton *et al.* (1984) found in their study, the importance of House Sparrows (from June onwards) and Starlings (from July) increased so that non-moorland species were contributing 39% of the Merlin diet in July and August. Both House Sparrows and Starlings are considered 'urban' species, and it is therefore significant that none of the Merlin nest sites in the Roberts and Jones study was more than 4km from the nearest hamlet or village. Roberts and Jones also record a decline in the fraction of these urban species in the Merlin diet from 1983 when the British population of both declined.
3. The variation between different months is statistically significant at the 99.9% level.

The variation in proportion of the two classes of prey species reflects the availability of prey through the season, with the 'preferred' open country species being scarce in the spring as they are migratory. A similar change in the ratio of woodland to open country as the breeding season progresses has been seen in studies in other parts of Great Britain, and has also been seen in Irish Merlins (Clarke and Scott (1994) for Northern Ireland, Fernández-Bellon and Lusby (2011a) for Eire).

The prey species list of Newton *et al.* (1984) was in good agreement with an earlier study in the same area (Newton *et al.* 1978) (Table 6), though the earlier study also found the remains of other bird species: Chiffchaff (*Phylloscopus collybita*), European Robin, Great Tit (*Parus major*), Marsh Tit (*Parus palustris*) and Willow Warbler.

Newton *et al.* (1984) also looked at the composition of the Merlin diet during the breeding season and confirmed the findings from North America that the number of fledglings (defined as birds with incompletely developed flight or tail feathers) in the diet increases during the time the Merlin pair are feeding their own nestlings (Table 9).

	April	**May**	**June**	**July**
	N (%)	**N (%)**	**N (%)**	**N (%)**
Full-grown birds[1]	229 (100)	483 (98.8)	367 (80.8)	181 (72.4)
Fledglings[2]	0 (0)	6 (1.2)	87 (19.2)	69 (27.6)

Table 9 Percentage of fledglings in the Merlin diet as the breeding season progresses. From Newton *et al.* (1984).

Notes:
1. Includes both adults and full-grown juveniles.
2. Includes 12 nestlings and two downy chicks of Eurasian Golden Plover and Red Grouse.

Table 9 clearly demonstrates the vulnerability of inexperienced fledglings to attack. However, what is equally striking is that the Merlin diet still mostly comprises adult birds even at this time.

Newton *et al.* (1984) also investigated the dietary difference between male and female Merlins, that there is such a difference being one of the theories regarding the distinct reverse sexual dimorphism of the species (and other falcons). The investigation was not straightforward as it was not possible to differentiate between prey remains which had been brought to the nest by the males and females, but Newton and co-workers were able to infer the difference by grouping the prey items recovered from plucking posts and nests into pre-egg laying, incubation/young chicks, and late nestling/post-fledge periods, as during the incubation/young chick phase only the male hunted. Within the groups which represent phases of the breeding cycle, the prey items were then divided into classes by weight. These data are shown in Table 10.

Breeding Phase	Number (Percentage) of items in the weight classes						
	≤10g	11-20g	21-30g	31-40g	61-80g	81-100g	>100g
Pre-laying	24 (7)	190 (55)	31 (9)	34 (10)	17 (5)	8 (2)	43 (12)
Incubation	12 (4)	243 (72)	22 (7)	51 (15)	8 (2)	0	1(<1)
Late Nestling/ Post-fledge	1(<1)	210 (53)	15 (4)	56 (14)	102 (26)	2(<1)	13 (3)

Table 10 Weight classes of prey taken during three phases of the breeding season. From Newton *et al.* (1984).

Note:
There were few birds in the weight class 41-60g in the study area and the remains of none were found among prey items.

From Table 10 it is clear that male Merlins primarily take prey weighing up to 40g (98% of all prey), with only one prey item (a Fieldfare, weight above 100g) being taken during the incubation phase. By contrast when the female Merlins were hunting the balance of prey weights shifted, with 19% above 40g in the pre-lay phase and 29% in the late nestling/post-fledge phase. Newton *et al.* (1984) note that these differences must have been due to differences in the prey of the sexes as they could not be entirely explained by seasonal changes in the prey availability.

In the study of Newton *et al.* (1978) on only three occasions the wings of moths were discovered, on one occasion the remains of a vole and on one other the remains of a bat. This low fraction of terrestrial mammal prey is consistent with other studies in Britain and also with studies in North America. However, Wiklund (2001) found significant numbers of rodents in the diet of Merlins in northern Sweden and considers that a possible reason for the disparity is that British studies of the species' diet

Juvenile Prairie Merlin with a dragonfly.
Laurie Koepke.

tended to collect remains at plucking posts, a technique which could seriously under-estimate the number of rodents killed.

Wiklund and Larsson (1994), studying Merlin distribution and nest sites in the Padjelanta and Stora Sjöfallet National Parks in northern Sweden found that Meadow Pipits and Northern Wheatear were the most important prey species, other prey being Bluethroat, Brambling, Chaffinch, Common Redpoll, Fieldfare, Lapland Bunting (Lapland Longspur), Redstart (*Phoenicurus phoenicurus*), Redwing, Reed Bunting, Snow Bunting and Yellow Wagtail (*Motacilla flava*). These species are probably indicative of the *F.c. aesalon* diet across much of Fennoscandia and European Russia west of the Urals, but an interesting study of Merlins in Belarus reported in Morozov *et al.* (2013) allowed the diet of the falcons to be analysed by habitat (see Table 11 overleaf).

Prey Species[1]	Percentage of Total Prey Items		
	Agricultural land, peat and pine bogs	**Pine bogs within forests**	**Pine bogs close to pasture and hayfields**
Common Snipe	-	11.1	-
Curlew	-	11.1	-
Great Snipe (*Gallinago media*)	-	3.7	-
Great Spotted Woodpecker (*Dendrocopos major*)	-	11.1	7.1
Great Tit	-	11.1	-
Greenfinch	4.5	-	-
House Martin	4.5	-	-
Meadow Pipit	13.7	7.4	-
Redwing	-	-	7.1
Skylark	13.7	-	-
Song Thrush	-	-	14.3
Starling	63.3	3.7	28.7
Tree Pipit	-	-	14.3
White Wagtail	-	3.7	7.1
Wood Sandpiper *(Tringa glareola)*	-	11.1	-
Other *Turdus* spp.	-	3.7	-
Phylloscopus spp.	-	-	7.1
Other *Passeriformes*	-	22.3	14.3

Table 11 Percentages of total prey in three habitats of Belarus. From Morozov *et al.* (2013).

Note:
1. The figures in the Table are for the major prey items. The Merlins also took Blue Tit, Goldfinch, Greenshank (*Tringa nebularia*), Jay (*Garrulus glandarius*), Lapwing, Moorhen (*Gallinula chloropus*), Barn Swallow, Treecreeper and Whitethroat (*Sylvia communis*).

Morozov *et al.* (2013) also give data on the diet of Merlins in other parts of their Russian range:

Middendorff's Grasshopper Warbler (*Locustella ochotensis*), an occasional prey item for Pacific Merlins.

On the Kola Peninsula, Merlin on the tundra took pipits, thrushes, Bluethroat, Common Redpoll, Northern Wheatear and Pied Wagtail. Falcons closer to wader colonies tended to take a greater fraction of those, one study noting Ruff (*Philomachus pugnax*) and Wood Sandpiper comprising 28.0% and 26.2% of the diet respectively. Closer to the coast Merlins were taking European Oystercatcher and (Ruddy) Turnstone (*Arenaria interpres*), and the chicks of Common Gull (Mew Gull – *Larus canus*) and Arctic Tern.

On the Kanin Peninsula the falcons took Lapland Bunting (Lapland Longspur), thrushes, Emberiza spp. and Bramblings.

On the Malozemel'skaya tundra, on the Russian mainland south of Kolguev Island, Red-throated Pipit (*Anthus cervinus*) comprised 23.8% of the diet, with Bluethroats and the ducklings of various species accounting for much of the remainder.

In the southern Yamal Peninsula, Lapland Bunting (Lapland Longspur) comprised about 45% of the diet with Red-throated Pipits accounting for a further 21%. Further east there is little data other than the general comment that pipits, Lapland Bunting (Lapland Longspur) and Common Tern form part of the diet, as do tundra rodents.

In the mountains of central Asia Merlins were taking a large number of local species, including such exotic species as Brown Accentor (*Prunella fulvescens*), Eversmann`s Redstart (*Phoenicurus erythronotus*), Himalayan Accentor (*Prunella himalayana*), Hodgson`s Rosy-Finch (*Leucosticte nemirocola*), Olivaceous Leaf-Warbler (*Phylloscopus griseolus*), and Red-fronted Finch (*Serinus pusillus*).

In the taiga, warblers provide a larger fraction of the diet while on the steppes (Naurzum State Nature Reserve, Kazakhstan) Merlins mainly took Northern Wheatear, Skylark, Stonechat (*Saxicola torquata*), Tawny Pipit (*Anthus campestris*) and White-winged Lark (*Melanocorypha leucoptera*), but also Booted Warbler (*Iduna caligata*), Quail (*Coturnix coturnix*), Whitethroat and local rodents.

In their study in the Altai-Sayan Mountains, Karyakin and Nikolenko (2009) did not carry out an exhaustive study of the Merlin diet, but noted that the most frequently predated species were 'open country' birds, including Larks, Anthus spp, Emberiza spp, Lanius spp, Motacilla spp, Oenanthe spp, Saxicola spp, Bluethroat, Black Redstart (*Phoenicurus ochruros*), Rock Sparrow (*Petronia petronia*) and Twite (*Carduelis flavirostris*).

Diet during Migration and at Winter Quarters

Prey during migration and at wintering sites is not well documented, though as it is clear that resident Merlins continue to take local, i.e. available, prey and it can be assumed that the eclectic nature of the Merlin breeding season prey means that the falcons will take whatever is available during migration flights and at a winter quarters.

Early North American studies (Allen and Peterson 1936) found that the diet of migrating birds included a large fraction of insects, the examination of 41 stomachs finding the remains of 115 dragonflies, two crickets and two grasshoppers, as well as 34 birds, two Red Bats (*Lasiurus borealis*) and a field mouse. Allen and Peterson saw a Merlin harassing a flock of Lesser Yellowlegs, but did not actually see a catch. Indeed, after many failed chases the observers noted that the Lesser Yellowlegs became inured to the Merlin and would feed close to the falcon as it sat preening on a post. Dekker (1988) and Cooper (1996) also observed migrating Merlins hunting dragonflies, the latter speculating whether the insects might be an important dietary item, particularly for juvenile falcons on their first migration who might find local avian prey more elusive targets.

Page and Whiteacre (1975) studied the winter diet (and hunting techniques: see *Chapter 5*) of Merlins at the Bolinas Lagoon, a 570ha estuary on the southern side of the Point Reyes Peninsula in central California, where the falcons were taking shorebirds and passerines (Table 12).

Prey Species[4]	Percentage of Total Prey Items
Least Sandpiper	37.2
Dunlin	35.9
Western Sandpiper (*Calidris mauri*)	8.6
Sanderling (*Calidris alba*)	6.0
Yellow-rumped Warbler	3.0
American Pipit[1]	2.3
Savannah Sparrow	2.3
Red-necked Phalarope[2]	1.3
Red-winged Blackbird	1.3
Red Phalarope (*Phalaropus fulicaria*)	0.3
House Finch (*Carpodacus mexicanus*)	0.3
Brown Creeper (*Certhia americana*) [3]	0.3
Western Bluebird (*Sialia mexicana*)	0.3
Ruby-crowned Kinglet	0.3
Song Sparrow	0.3

Table 12 Diet of wintering Merlin in California. From Page and Whiteacre (1975).

Notes:

1. Page and Whiteacre quote Water Pipit (*Anthus spinoletta*), but, as noted above (Table 2: Note 1) it is now considered that the American Pipit is a distinct species from the European species.
2. Page and Whiteacre call this species the Northern Phalarope, but the name has since been changed.
3. Page and Whiteacre give the scientific name of this species as *Certhia familiaris*, but the designation has since been changed.
4. Page and Whiteacre also record an estimated four California Voles (*Microtus californicus*) and an unidentified lizard in the Merlin diet.

Urban Merlins in Saskatoon, where the birds are resident, allowed Warkentin and Oliphant (1990) to study not only the change in diet from breeding Merlins, but to study changes in the birds' hunting ranges and techniques (see below). The researchers fitted some trapped birds with radio transmitters so as to be able to monitor their movements, while others were fitted with coloured leg streamers after trapping so they could be identified (**Box 7**). The results of the study showed a significant overlap between the summer and winter diets, reinforcing the suggestion above that the Merlin winter diet is largely based on the availability of prey. In summer House Sparrows form the greater majority of prey (Table 4). That is still the case in winter, but Bohemian Waxwings, which do not breed in large numbers in the vicinity of Saskatoon, but do overwinter in/close to the city having migrated south from their breeding grounds, formed a larger fraction of the diet than in summer (**Box 8**).

Box 7
The Merlins were trapped using a dho-gaza mesh trap which was baited with two tethered House Sparrows. Use of live trap bait will surprise many, particularly in Britain where such a technique is outlawed, but at the time use of House Sparrows did not require regulatory permission as *Passer domesticus* is an introduced species to North America. Warkentin and Oliphant note that the usual hunting technique of the Merlin is to surprise its prey and flush it into the air where it is captured. The tethered sparrows were therefore unlikely to be killed and, indeed, only four were killed in 248 exposures (a 1.6% fatality rate).

Box 8
Servheen (1985) suggests that the increase in winter Bohemian Waxwing populations in the cities of Canada and northern States of the USA is perhaps a response to the increased planting of ornamental fruit trees such as Sorbus spp. and Malus spp. which offer the birds a reliable food resource. The increased Bohemian Waxwing population in these areas has been accompanied by an increase in the urban Merlin population, and also by a northward expansion of the Merlin's wintering range, which James *et al.* (1987a) suggest might be due to an increased continental population following the decline caused by organochlorine contamination.

What is perhaps most notable about the urban winter diet is the reduction in prey species, a measure of the reduced number of species populating the city in winter (Table 13), though it is also the case that fewer prey remains were collected in winter. What is also interesting is the increase in the mammal fraction of the diet.

Prey Species	Percentage of total prey items	Percentage of total biomass
House Sparrow	72.2	66.2
Bohemian Waxwing	17.3	26.3
Meadow Vole	6.0	5.0
Rock Dove	2.2	1.7
Common Redpoll	1.5	0.5
Pine Siskin	0.7	0.3

Table 13 Winter prey of urban *F.c. richardsonii*. From Warkentin and Oliphant (1990).

The importance of Bohemian Waxwings to wintering urban Merlins was confirmed by Servheen (1985) who studied the falcons in Missoula, a city in west-central Montana with 36,000 inhabitants at the time of study which, remarkably, was home to all three Merlin sub-species during the winter period. All three hunted the waxwings.

In Britain Dickson (1988) studied wintering Merlins over a 19 year period (1965-1984) in Galloway, south-west Scotland where the falcons were hunting an area comprising moorland, lowland cultivated country and coastal sites. In this study the Merlins were primarily hunting Skylark and finches (including Common Redpoll, Greenfinch, Linnet and Twite), but also waders (Dunlin, Lapwing, Redshank, Ringed Plover and (Ruddy) Turnstone), as well as Blackbird, Pied Wagtail, Redwing and Reed Bunting. Dickson also monitored the number of times male and female Merlins attacked prey, and the success rate of attacks. In the case of both males and females the success rate was the same, at about 10% (see **Box 9** overleaf). (Note: Dickson defined 'blue' and 'brown' Merlins, the present author making the assumption that the colours were sex related, i.e. male = blue, female = brown. However, Dickson noted that brown Merlins were more likely to attack larger birds, and since the possibility of adult males taking smaller prey than first year (brown) males is minimal, the colour/sex assumption seems reasonable, as Dickson himself agrees.) In a further analysis, Dickson (2000) extended his analysis of the Merlin

Box 9
In *The Art and Practice of Hawking,* written by E.B. Michell, Michell states that a Merlin called *Sis,* a falconry bird, achieved a 91% kill rate against larks (59 of 61) including 'the extraordinary score of forty-one out of forty-two successful flights, the one miss being a ringer at which she was thrown off when the head of another lark was hardly down her throat – before she had shaken herself, or had time to look around.' This seems a remarkable success rate for a falconry bird, but the veracity is hardly in doubt. Edward Blair Michell (1843-1926) was Oxford-born and Oxford educated, and became a barrister. He was a linguist, being fluent in French and Siamese, and a sportsman, winning several important rowing championships. He was also an expert falconer, particularly knowledgeable regarding the Merlin. His book is still regarded as one of the foremost books on falconry. While the larking season saw healthy Merlins flown against juvenile larks, in general falconry birds are likely to be less fit than wild birds making the success rate astounding. For further consideration of the success rate of hunting wild Merlins see *Chapter 5.*

winter diet by another eight years (1992-2000). In those years 'blue' Merlins preferentially attacked Linnets and Twites (64.7% of attacks) and Chaffinches (17.6%), with Skylarks being the target on only 8.9% of occasions. This compares to the earlier data (Dickson 1988) in which the combined finch fraction was 31.6%, with Skylarks accounting for 21%. For 'brown' Merlins the rate of attacks on Linnet and Twite had also increased (58.6% cf. 11.6% for all finches in the earlier study), though in this case the Skylark fraction had also increased (32.9% cf. 21%).

Migrating Eurasian Merlins are believed to feed on the Barn Swallows that are also migrating, as well as on local species of the areas through which they pass or in which they winter. As an example, in a study of migrating and wintering Merlins in ex-Soviet Georgia (Abuladze 2013), examination of the stomach contents of birds shot by poachers revealed 23 bird species, chiefly passerines. However, Abuladze notes that as observations suggested the Merlin were only successful in 26% of their forays, the number of species which might actually comprise the potential diet could be much higher. The most prevalent species in the wintering diet of the Georgian birds was the Chaffinch, followed by House Sparrows, and three species which would not normally be seen by Merlins at their breeding grounds – Crested Lark (*Galerida cristata*), Quail and Red-backed Shrike (*Lanius collurio*) – again indicating the falcon's willingness to take whatever it could get its talons on. Morozov *et al.* (2013) also include data which emphasises the eclectic nature of the diet of both migrating and wintering Merlins, noting that local rodents and lizards were added to a diet of local bird species.

Juvenile Lapland Bunting (Lapland Longspur). The species forms part of the Merlin diet wherever the ranges of the two overlap.

A further example of the willingness to take local prey is noted by Hilty (2002) who, considering the diet of Merlins wintering in Venezuela, found that the falcons were feeding on Dickcissel (*Spiza americana*) flocks which were themselves overwintering at Los Llanos (meaning 'the plains'), a huge area of seasonally flooded tropical grassland lying at the eastern foot of the Andes which stretches across northern Colombia and western Venezuela. As Dickcissels breed in the southern states of the USA it is likely that the species provides a food resource for both migrating and wintering Merlins.

Raim *et al.* (1989) studied the diet of a female Merlin trapped and fitted with a radio tag during a stopover on the island of Loggerhead Key in the Gulf of Mexico, 126km west of Key West, Florida. During the seven days the falcon spent on the island before resuming its migration north to the mainland, the Merlin preyed on any available local avian species weighing 10-40g. Although some captured species could not be identified, the female showed a particular enthusiasm for Yellow-billed Cuckoo (*Coccyzus americanus*), Tree Swallow, Indigo Bunting (*Passerina cyanea*),

Female Snow Bunting. The bunting is another species which forms part of the Merlin diet wherever the ranges of the two overlap.

Ovenbird (*Seiurus aurocapillus*), Palm Warbler (*Dendroica palmarum*) and Black-throated Blue Warbler (*Dendroica caerulescens*).

Further evidence for the opportunistic way in which wintering Merlins exploit local resources is offered by Rodriguez-Durán and Lewis (1985) who studied the birds overwintering in western Puerto Rico across two successive winters (Merlin are the most abundant resident winter raptor on the island – see Rivera-Milán 1995). Rodriguez-Durán and Lewis observed Merlins taking Sooty Mustached Bats (*Pteronotus quadridens*), an insectivorous species, as they emerged from their roosting caves. One Merlin was noted to take an average of 5.1 bats/day, and the study estimated that a total of 1800 bats (1.5% of the population) was taken during the first winter. By the end of the second winter, the bats had learned to avoid the most vulnerable of the flyways they used on their outward journey.

In another study, on East Ship Island, which lies off the Mississippi coast, Chavez-Ramirez *et al.* (1994) noted that the appearance of Merlins (and Peregrine Falcons) on the islands coincided with that of migrating songbirds, and that there was a positive correlation between Merlin numbers and songbird numbers. Migrating Merlins therefore exploited both local and migratory populations, to fuel their journeys.

Also interesting are the observations of Odin (1992) who watched Merlins (a total of 34 observations in a 57 day period) hunting passerines above the sea off Kent, south-east England. Odin considered the Merlins were not returning migrants, but falcons wintering locally: it was not clear whether the captured passerines were migrants, though the observations (from late February to mid-April) suggest that is possible. The Merlins were often successful in their hunting suggesting they were taking advantage of the lack of cover available to the passerines (and, potentially, their exhausted state). Equally interesting is the observation of Dean (1988) who saw a wintering juvenile Merlin returning from the sea off the Cumbrian (north-west England) coast clearing flying with difficulty because of the prey it was carrying. Dean believes the Merlin was first seen over 1km from shore. When it arrived at the beach it was seen to be carrying a Leach's Storm-petrel (*Oceanodroma leucorhoa*). The Merlin ate for 30 minutes, then departed leaving only the Storm-petrel's wings, legs and upper mandible.

Food Caching

The reasons for prey caching appear straightforward – they provide a larder for food caught in abundance when conditions or prey availability are favourable to be used when the weather prevents hunting, or when prey density declines. But studies carried out on other species suggest that more subtle reasoning is at work. From a study on Eurasian Kestrels, Masman *et al.* (1986) in Holland provided interesting data on the weights of voles eaten, fed to chicks or cached which indicated that adult birds tended to consume smaller voles, feeding larger specimens to the chicks and caching the largest. In their study of the energy intake of the Kestrels Masman *et al.* (1986) also noted that it was clear that the birds were heavily dependent on cached food during days when the weather prevented hunting, a fact which probably explains the reason for caching larger prey. No similar studies appear to have been carried out for Merlins, but the similarity of size of the two small falcons, the food intake requirements of their nestlings, and weight spectrum of their respective prey allow the contention that a similar strategy may well be at work in Merlins.

Merlins are known to cache prey during both the breeding season and in winter. Greaves (1968) observed the caching of a Meadow Pipit by a female *F.c. aesalon* in north-east Scotland in August ('about 50 feet from the spot, lying in a sleeping bag wrapped in a mudstained green fly-sheet (having spent the night there)') and records another instance of similar behaviour in Ireland seen by another observer, also in August. Greaves also records another observer watching a female who was brooding chicks leaving them to retrieve cached prey after her partner had not provided fresh food for about two hours. Sperber and Sperber (1963) also record instances where female Merlins in Norway received food from her mate, but did not take it to the nest, returning to her brood without food, then later leaving the nest again and returning with food even though there had been no prey delivery by the male. Clearly the female was caching food and then retrieving it later, presumably gauging chick hunger in order to decide on feeding times.

In a study of urban-dwelling Merlins in Saskatoon, Oliphant and Thompson (1976) noted a male Merlin caching food during all phases of the breeding season, caching initiated by the female's refusal to take food when her partner offered it. The male, having prepared the prey by removing the head, legs/feet and wings and plucking, or partially plucking, it would fly close to the nest and call the female. If the female failed to respond, the male would then fly to a different tree and cache the food. Males were also seen caching prey which they had partially consumed.

Oliphant and Thompson saw only one incident of a female food caching during the breeding season (incubation phase): later the same day, when the male had not brought fresh prey, the female retrieved and ate the cached food. Oliphant and Thompson saw food caching at all times of day except early morning, suggesting that the female Merlin was always hungry at that time. Caches were always in trees, the Merlins tending to use different trees rather than a particular one, and using elms less often than spruce once the former had started to shed their leaves, suggesting that concealment was an issue. However, while all observed caches were in trees, Oliphant and Thompson note the work of another researcher in southern Alberta who had observed ground caches, as seen in many British caching observations.

Pitcher *et al.* (1979) record winter caching in another North American city (Sheridan, Wyoming), observing both male (a juvenile bird) and female Merlins caching or retrieving cached food, while Warkentin and Oliphant (1985) record winter caching in their study in Saskatoon.

Greaves (1968), records seeing a female Merlin caching a Meadow Pipit, and also, as noted, records incidents of others who had observed caching. In the other cases Greaves says the Merlins readily retrieved their caches, though he does not say whether the cache he observed was retrieved, but Oliphant and Thompson (1976) note that the male Merlin they were watching had to search for the food item. Warkentin and Oliphant (1985) also observed Merlins searching for cached food, and noted apparently deliberate search patterns, the falcon starting high in a tree and hopping purposefully down from branch to branch searching until it reached the ground. Such search patterns suggest that Merlins are not good at recalling where they have cached. The Saskatoon studies might offer an explanation for that as Oliphant and Thompson (1976) noted that the Merlins rarely used the same tree for caching during the breeding season, while Warkentin and Oliphant (1985) noted that the same two trees were consistently used in winter. This suggests that during the breeding season theft of cached food might be a problem, the pressure on adult birds (not only other Merlins, but also the significant corvid breeding population of the city) to provide food for nestlings being high. The use of many trees for caching might then mean the Merlins needing to employ search patterns to find their caches, particularly if retrieval is several hours after storage. In his study of ground nesting (and ground caching) Merlins in north-east England, Wright (2005) notes that females occasionally retrieved prey from where their mates had cached it, even if the female had not been present or nearby when the male had secreted it. This suggests either the use of preferred sites, or that female Merlins, perhaps away from the nest to hunt or to preen, had actually observed the caching, but from such a distance that the human observer was unaware and so was amazed by her apparently telepathic abilities.

One interesting aspect of caching was noted by Warkentin and Oliphant (1985) who observed one Merlin hunting immediately after having cached food. Winter nights in Saskatoon are long and cold and this behaviour seems to indicate a deliberate strategy of preparing food either for later the same day (i.e. before nightfall) or for early morning consumption (i.e. the following day). Winter caching would appear to allow the bird to profit from an evening meal at a time when night-time requires enduring the long, cold hours of darkness with no opportunity to hunt. Coupled with the possibility of poor weather the next day which might make hunting difficult, this would make winter caching a survival strategy.

In addition, Rijnsdorp *et al.* (1981) suggested, from a study of the Eurasian Kestrel, that caching an evening meal allowed the bird to hunt most efficiently by reducing body weight during the day, quoting a figure of 7% of daily energy expenditure, if a meal was consumed just before nightfall. However, the observation of Rijnsdorp and co-workers was based on a single male Kestrel eating during summer evenings and so may be explained by the need to feed chicks during the day.

Warkentin and Oliphant suggest the retrieved cache would have been frozen, but saw no evidence of any different feeding behaviour. While they note that the caloric value of a frozen food cache will be lower than that of a fresh kill, for an individual bird that might make the difference between being able, or not being able, to have the energy to hunt.

Hunting Strategies

Trimble (1975) notes that the Merlin's ability to twist and turn is superior to that of other falcons (though he had not, perhaps, seen a European Hobby hunt), and almost the equal of Accipiters. The Merlin's size and flying ability made it popular as a falconry bird – not necessarily only with ladies despite the claims of the Book of St. Albans (see *Chapter 1*) though there is no doubt that ladies of the European court did favour the birds. The preferred sport was the 'ringing' flight in pursuit of larks. In this, the Merlin was released in pursuit of a lark which had been startled into flight by the falconer or his dog (**Box 10**). If the lark was far from cover, then being slower than the Merlin in level flight it would fly upwards as it could gain height faster than the falcon, about 25% of the mass of a Skylark being flight muscle, which gives them good climbing ability. In classic flights the lark would make tight circles (rings) as it climbed, the Merlin following, usually on a wider circle to contain the prey, until, occasionally, both birds were out of sight having climbed several hundred metres. The Merlin would then return with or without its quarry. Successful ringing attacks involved the Merlin outflying the prey, either closing in on and ultimately catching, the tiring prey from below, or managing to climb above the exhausted lark and stooping onto it. Merlins were usually successful against juvenile birds, but against adults the falcon might well become exhausted first and return to the falconer empty-taloned. The classic ringing flight was observed by Servheen (1985) in an attack by a Merlin on a Bohemian Waxwing in an urban environment in Montana, though Servheen does not actually use the term. He noted that the Merlin pursued the Waxwing at a distance of about 6m into a small stand of conifers. As the Waxwing was unable to find protection – presumably having insufficient time to land – it 'left the trees. The Waxwing then began to fly upward in a spiral with the Merlin flying below in large circles to gain altitude.' In this case the Waxwing escaped. Interestingly, Dunlin, a frequent winter prey species for Merlins, also have good climbing ability (Hedenström and Rosén 2001), but rarely attempt to fly upwards to avoid a Merlin attack, though Dunlin have been observed to attempt outclimbing a predator (see, for instance, Lima 1993).

However, although ringing flights are seen in the wild, they are not the usual hunting technique. Merlins usually hunt from a perch, chosen to offer a wide view across the bird's hunting range, though the bird will quarter the ground, making fast flights either close to the ground,

Box 10
In his classic work on falconry, particularly Merlin falconry, Michell (see *Chapter 3: Box 9*) notes that 'larks, for hawking purposes', may be divided into three kinds'. These are the 'ground lark, generally deep in moult' which flies low and fast in an attempt to get to cover. Next, the 'mounting lark' which attempts to rise so that the Merlin cannot get above it to stoop. If it succeeds and gains enough distance, it will make for cover. If it fails, it will become the third kind, the 'ringer'. To these must be added a fourth category, those birds which sit tight on the ground in the hope the Merlin will not notice them. For falconry purposes these offer little sport, but from the lark's point of view, this may be a strategy which offers the better chance of survival. Michell also quotes a hunting success rate for an individual falconry Merlin which is much higher (see *Chapter 3: Box 9*) than the rates quoted in this Chapter from studies of hunting Merlins.

following its undulations, in open country, or below the tree tops in more boreal areas in the hope of being able to make a surprise attack. Merlins will also stoop in the manner of Peregrines. Long stoops are rare, though a series of short stoops may be utilised, seemingly as a way of gathering speed or adding surprise to an attack on prey which is proving difficult to overhaul in straight flight. In his study of *F.c. aesalon* in east Scotland, Cresswell (1996) noted that in winter hunts the Merlins used surprise attacks 62.2% of time, with ringing flights 29.9% of the time, but noted that ringing often followed a surprise attack which failed to take the prey. Cresswell noted that surprise attacks were mainly launched from a perch (83.3% of all attacks) at prey which was usually on the ground (78.3%) and close to the Merlin (60.2% were at <100m, 37.0% at 101-500m). Overall, Merlins spent 83% of their time perched and 7% feeding, the remaining time hunting. One surprise attack which did not result in a ringing flight was seen in Scotland by MacIntyre (1936) who watched a Merlin chase a lark (Macintyre does not say which species, but probably a Skylark, though at the time Meadow Pipits were often called larks) which chose not to try outclimbing the falcon, but dived beneath a cow, circled over the cow's back, and dived beneath the cow again. The Merlin followed, but could not keep up on the second circuit and rose in the air. The lark, now with a little time, settled close the cow's hoof where it was safe from attack.

Sodhi *et al.* (1991a) record forms of both the main 'open country' techniques being utilised by urban-dwelling Merlins in Saskatoon. Attacks from perches represented 58% of summer attacks and 95% of those in winter: the winter data is consistent with that of Cresswell (1996) and suggests that Merlins are organising their time budget to minimise energy expenditure at a time when bodily heat loss is maximal, though in a study of wintering Merlins in south-west Scotland Dickson (1988) found that the falcons hunted primarily by 'low flight' in which the falcon flew fast over the ground at a height of about 1m in an effort to surprise prey into the air. This mode, which is a high energy hunt, represented 49% of observed hunts. Dickson also noted perch hunting (32% of total hunts), the remaining 19% of hunts being higher flights in which the falcon searched the ground for prey. In a separate study, Dickson (1996) again mentions low flying as a hunting technique, but as a form of perch hunting, implying that such flights are against a predetermined prey observed from a distance of 50-300m which, from a point of view of the bird's winter energetics, would make more sense. Dickson (1996) also mentions 'bounce tactics' in which the Merlin makes a downward thrust towards avian prey which has crouched on the ground as a means of defence, the 'bounce' being a technique to force the prey into the air. Dickson (1996) also notes that Merlins will occasionally stalk prey, walking across the ground using vegetation as cover until close enough to pounce. Dickson says his observations represented only 1% of hunts and involved stalks of only 2m or so, but Fleet (1993) observed a Merlin walk and hop over 30m in a ground stalk of a Dunlin flock. The Merlin then took to the air for the last 30m to the flock where it successfully caught a bird. As strange as stalking on foot is the record of Craib (1994) who looked over a wall to catch a glimpse of a Merlin seen landing on the ground. Craib saw nothing until the falcon emerged from a rabbit burrow and flew away after catching sight of him. A few seconds later a juvenile Starling, which the Merlin had clearly pursued into the burrow, emerged. Though apparently carrying a wing injury the Starling flew off. Craib looked into the burrow and was surprised to see a second Starling: Craib was able to reach and retrieve this bird which flew off uninjured.

In the Saskatoon study Sodhi *et al.* (1991a) observed attacks from high perches which were either direct flights to the prey that had been spotted, or involved a glide towards the ground, followed by a flight close to ground level. Sodhi and co-workers also observed an attack in which a Merlin landed in a tree in which there were several perched House Sparrows, then hopped downwards from branch to branch until it was close enough to force a sparrow to flee and then pursued it. Low level flights were observed to be used 37% of the time in summer, 5% in winter. In these cases the Merlin used the same technique as in open country, but utilised streets as flyways, often startling prey from ground cover as well as from trees, and also taking sparrows which were flushed from close to, or beneath, parked cars. The Merlins also took prey opportunistically if they had been flushed by, for instance, a dog or another raptor (**Box 11**). Sodhi *et al.* (1991a) also noted three other hunting techniques which comprised 5% of summer attacks. The first of these was ringing which was observed against both solitary birds and

Box 11
Other examples of opportunistic hunting have also been recorded. Jenkins (1972) records flushing a Horned Lark (Shore Lark) from the roadside in Utah and seeing a Merlin take it. Jenkins believed that the falcon had been deliberately flying above and behind his car waiting for such an opportunity. Kenyon (1942) also noted up to six Merlins routinely following trains in Arizona watching for small birds to be flushed.

flocks. The second involved a straightforward speed duel between the Merlin and prey such as swallows and juncos which attempted to outfly or outmanoeuvre the predator. These flights, which are also occasionally seen in open country, can involve the Merlin climbing to gain acceleration in a stoop, but may be level flight chases, continuing until either the prey is caught or the Merlin disengages. The third technique observed were stoops from altitude, in the manner of larger falcons. Sodhi and co-workers also observed attacks in open country when the Merlin started from a ground perch and attacked prey from below, forcing it upwards and away from cover, a variation of ringing flights. One interesting aspect of the study of the urban-breeding Merlins of Saskatoon is that they rarely hunt close to their nests (unpublished data quoted in Sodhi *et al.* (1990) which suggests that male Merlins usually hunt at a distance from their nests, but may make opportunistic attacks close to the nest) though the reason for this is not suggested. This is consistent with a study of Merlin in Scotland (Rebecca *et al.* 1990) which collected the rings of Golden Plover and Lapwing chicks from three nests (of five under study during 1986-1988) which had been predated by Merlins and fed to their brood. The chicks were aged 4-32 days, (with a mean age of 10.3±7.1 days). The mean distance the Merlins flew to take the chicks was 3.4±1.1km (range 2.0-5.6km).

Dickson (2005) observed a female Merlin hunting when the nestlings were 14-15 days old: the female caught a Meadow Pipit which she fed to the young. However, the male brought six prey items in the next three hours and the female cached two of these. Overall the male seemed a good provider, so the reason for the female to hunt was not clear. The female always hunted within 100-200m of the nest and continued to hunt through to the time the young fledged, often giving her catch to the young, but occasionally taking moths which she consumed. The reason for this behaviour is not apparent, though the fact that the female ate the insects herself suggests she was supplementing her diet, as well as that of her brood, despite her mate apparently being a good provider.

Although not specifically mentioned by Sodhi *et al.* (1991a) Merlins must also occasionally seize prey from the ground as their diet is known to include terrestrial mammals (in small fractions of the total diet) and also lizards. Fox (1964) also notes that after some chases Merlins will force their prey onto the ground and take it from there, and there are several examples of falcons wintering at the coast knocking prey into, or forcing prey to dive into, the sea and then collecting the prey from the water (see below). Fox (1964) also notes that he had seen American Goldfinches feeding on thistles being oblivious to a fast approaching Merlin so that the finch was taken from the plant before it could become airborne, a form of ground attack.

One further hunting 'technique' must also be mentioned. Uttendörfer (1952) notes that Merlins will occasionally mimic the flight pattern of their prey in order to close in on it before launching an attack. This technique is not mentioned by Sodhi *et al.* (1991a), but Rudebeck (1951), whose observations of hunting Merlins include excellent descriptions of the flight pattern of Merlins (as noted above), mentions a falcon which was flying 'with half folded wings and very rapidly vibrating wing-beats. In flight it resembled a swallow and a thrush.' Rudebeck goes on to say of one hunting Merlin that 'in a quite remarkable way it resembled a swallow, in outline as well as in behaviour. Nearby and at a somewhat lower altitude there were several swallows: they kept quiet and silent and apparently did not recognize the Merlin as a source of danger.' Rudebeck then quotes another observer who twice noted a Merlin 'flying in a very peculiar manner in the midst of a flock of swallows which showed no fear at all'. On each occasion the Merlin was able to take a swallow. This clearly leaves a question – does the flight pattern of the Merlin resemble that of its occasional prey, or does the Merlin deliberately mimic flight patterns in order to get close to prey?

Sodhi *et al.* (1991a) consider the difference between the observed techniques adopted in summer and winter attacks probably derive from differences in prey behaviour in the two seasons, e.g. the prevalence of flocking, or from habitat usage, or energetic constraints, or a combination of these. Sodhi and co-workers also looked at the success of attacks, finding that in summer 27% of perched attacks and 21% of cruise attacks were successful, though the difference between the success rates was not statistically significant. In winter the success rates dropped to 11% and 8% respectively. Sodhi *et al.* consider the difference probably results from the increased vulnerability of recently fledged prey species in summer, which is certainly consistent with the increased take of fledglings during that period (Sodhi and Oliphant 1993). It is also probable that breeding Merlins are good hunters, having proved themselves by surviving the previous winter: juvenile Merlins in their first winter are more likely to succumb to starvation as their hunting techniques are not honed sufficiently to find enough prey to stay healthy. Although resident Merlins

– which include urban birds – have the advantage of familiarity of territory, migrating Merlins may be less familiar with overwintering quarters resulting in lower attack success, as lower success rates are also seen in non-urban wintering birds, as we note below.

The observations of Rudebeck (1951) at Falsterbo in Sweden include a fine description of the Merlin's flight – 'The flight seems wild and vigorous although the bird is so small. The Merlin possesses an extraordinary capacity to turn round and make sudden zig-zags at great speed. It reacts with the utmost swiftness and is full of vitality: it may almost seem to have a nervous vibration in its movements. The latter is sometimes so pronounced that one has the impression of an entirely new flying technique, and the bird takes on a changed appearance. The tips of the wings are directed backwards, the 'wing knuckles' [i.e. the wrists] become still more pronounced and lie close to the sides of the body. The wings are thus not fully extended but are moved in a similar way to that of small passerine birds. The wing-beats become trembling and extremely rapid. Gliding on outspread wings is of course impossible in this position, but the wings can be kept still for short moments during which the bird glides ahead by its own momentum as many small birds do. The tail is narrowed and closed.' Rudebeck (1951) also logged the outcome of observed hunts from which he was able to conclude that Merlins were successful in only 4.5% of attacks (7 of 155). Rudebeck quotes the notes he made during a handful of the attacks, all but one of these entries being in September/October, consistent with his having observed migrating Merlins at Falsterbo. However, without information on the majority of attacks it is not possible to be definitive about when the bulk of the attacks were observed, though the success rate is lower than that found for wintering urban-dwelling Merlins by Warkentin and Oliphant (1990), who quote a value of 12.9% (comprising a success rate of 8.6% for first winter birds with a rate of 13.8% for more experienced adults), and by Sodhi *et al.* (1991a), who quote a value of 11%. These higher values probably reflect the higher abundance of prey in the city which allows more opportunities for attacks.

In 343 attacks by Merlins wintering in central California observed by Page and Whiteacre (1975) 278 were against prey on the ground and 16 at perched birds, suggesting surprise attack is the most likely to produce a result. The observed success rate of the attacks was 12.8%, more consistent with the data from the urban Merlin study. One interesting aspect of the observations of Page and Whiteacre was that of 17 successful attacks on wader flocks on the ground, three were against birds which hesitated momentarily when the remaining flock took off, and two were against birds 1-3m away from the main flock, but the remaining 12 attacks took birds from the flock. At first glance this seems at variance with the idea that flocking is a strategy to avoid predation, but such is not the case. For the birds in the five individual attacks, the outcome was 100% fatal, while for the 12 birds captured in flocks the outcome was dependent on the size of the flock, and as flocks of 5,000 birds are not unusual, the odds of the Merlin choosing one individual would seem to be very low. We return to consideration of flock size as a deterrent below.

Buchanan *et al.* (1988) noted a much higher success rate (22.5%) for Merlins attacking Dunlin flocks at estuaries in western Washington state, though there is a need for caution when comparing this figure with those given above as Buchanan and co-workers define a hunting flight as one which might involve a number of individual capture attempts (attacks), the latter having a much lower individual success rate (about 4.9%), more consistent with other observations. In a later study, Buchanan (1996) found that Merlins were more successful at capturing Dunlin at estuaries than at beaches where the waders occasionally roosted at high tide. Since the Dunlin were feeding at the estuarine sites they were presumably more vulnerable there as a fraction of the 'roosting' birds would always be watching for predators. (As a digression, the study of Buchanan (1996) was of both Peregrines and Merlins, each falcon attacking Dunlin: the success rate of the larger falcon was approximately double that of the smaller species.)

Buchanan *et al.* (1988) suggest that Merlins use seven different hunting techniques, though these are essentially similar to the smaller number of techniques noted by other observers, being listed as high-angle stoops against Dunlin flocks; stoops against single Dunlin; low angle stoops against flocks or individuals; stealth attacks, made close to the ground, against a flock; similar stealth attacks against an individual; ringing attacks; and horizontal passes through a flock, used when other attacks had failed. Of these, the final attack, and low-angle stoops, invariably failed, while the other techniques were not statistically different in terms of success rate (though stealth attacks were the most successful). In their observations Buchanan and co-workers noted that 54% of hunting attempts lasted less than one minute, with only 10% lasting more than five minutes (Fig. 2). The data from which Fig. 2 was constructed showed a statistically significant difference between the number of hunts lasting less than a minute to those lasting two minutes at the 99.9% confidence level, i.e. Merlins are very much more likely to hunt for periods of less than one minute. However, the success rate for hunts shows no statistically significant difference across the spectrum of hunting periods, i.e. although Merlins prefer short hunts, such hunts are no more successful than longer ones.

The data of Fig. 2 are consistent with that of Cresswell (1996) for Merlins hunting Dunlin, Redshank and Skylarks in Scotland – though the

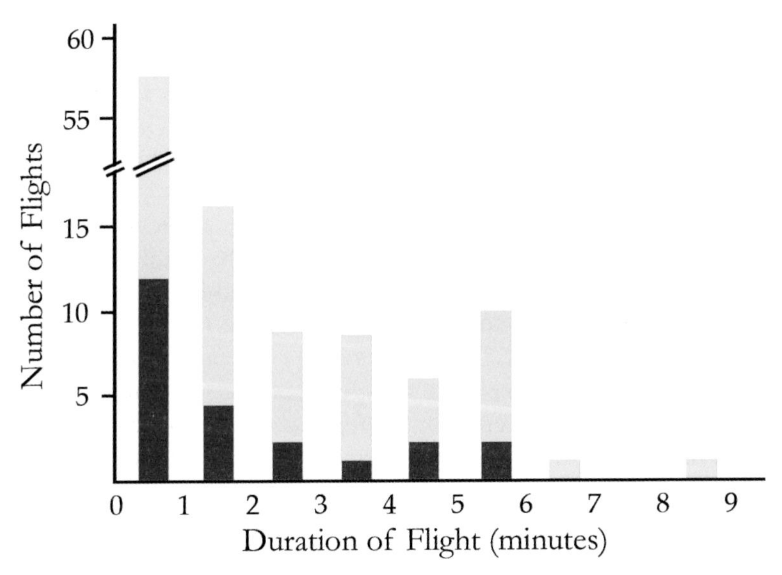

Figure 2
Duration of hunting flights by Merlins in pursuit of Dunlin. Red bars represent successful hunts. Note that these data do not differentiate between the various hunting techniques adopted by the Merlins, a hunting flight being defined solely as one which involved an attempt, or several attempts, at capture of a prey individual or several attempts at more than one individual. Bars represent times within periods <1 minute, 1-2 minutes etc.
Redrawn from Buchanan et al. (1988).

percentages for hunt times longer than one minute change between the two studies, the essentials remain the same, i.e. hunts lasting less than one minute are the bulk of all hunts (50% in each case) with a significant tailing off thereafter. Within the one minute hunt, Cresswell (1996) broke down the actual timings of observed attacks on Skylarks and found that the attack itself lasted less than one second in 41.4% of attacks. For the remaining attacks, the initial surprise failed, and a chase ensued, these lasting 1-10s on 25.4% of occasions; 11-30s on 9.5%; 31-60s on 6.3%; and more than 60s on 17.4% of occasions. Cresswell noted that many of the attacks involved no final stoop, but that longer, ringing flights ended with stoops on 50% of occasions. While it might be assumed that these would be single stoops, which either succeeded or failed, the hunt terminating at that point, Cresswell found that only 38.1% of ringing flights which ended with a stoop involved a single stoop, 30.2% involving 2-5 stoops and 31.8% involving six or more stoops.

Dekker (1998) also observed wintering Merlin preying on Dunlin flocks, in this case at Boundary Bay, British Colombia. He noted a high success rate (17.2%), but had a sample size of only 29 attacks. Interestingly, while all these began as surprise attacks, the five successful hunts all developed into pursuits. Dekker also confirmed an observation by Buchanan (1996) that the Dunlin remained airborne when their intertidal feeding grounds were underwater. Buchanan noted that when at sea the Dunlin engaged in synchronised flights in the troughs between waves. Both Buchanan and Dekker (1998) believed the tactic was an anti-predator defence, and this seems confirmed in another study (Dekker and Ydenburg 2004) which noted that the kill rate of Dunlins was highest just after the tide began to ebb and the waders returned to feed: that period, when the waders are closest to shoreline vegetation and so most vulnerable to surprise attack, is the most dangerous.

Dekker (1988) studied Merlin and Peregrine attacks on shorebirds and passerines in an open area of country around Beaverhills Lake in Alberta, Canada in both spring and autumn over a number of years. The autumn season corresponded to migration and the Merlin population would have included a number of juvenile birds whose hunting abilities would have been improving, but the spring season included the start of breeding and so would have involved experienced birds. Results were also loaded towards spring by the number of observation days in the two periods. But despite this, the Merlin hunting success rate was relatively low, at 12.4%. More interesting were Dekker's observations on hunting techniques and the success of male and female Merlins. Dekker noted that almost all attacks used the surprise technique (72.3% of all attacks) and postulated that Merlins use persistent hunting only if surprise attacks are unproductive, further surmising that this lack of productivity might result from an inexperienced falcon, prey scarcity, poor habitat or the success of defensive flocking by the prey. On the sex difference in attack success, Dekker found that male Merlins were more successful than female and juvenile birds combined, a difference that was statistically significant at 99.99% confidence.

In a study of *F.c. aesalon* wintering near the Tyninghame estuary in east Lothian, Scotland, Cresswell (1996) noted that the Merlins also mainly used the surprise attack strategy favoured by their American cousins. Cresswell's study was of three raptors that worked the Scottish site, the Merlin, Eurasian Sparrowhawk and Peregrine, and the differences between them are interesting. Cresswell considered the position of the raptor when an attack started. For the Peregrine the falcon was usually in the air when it started an attack (80% of all attacks – see Table 14), whereas Merlins started 82% of attacks from a perch or the ground. (The starting position of the Sparrowhawk was rarely seen and so was not given as a comparison.) For the Peregrine, in 60% of attacks the prey was not in flight (perched or on the ground) at the start of the attack (Table 14), while for the Merlin the prey was not in flight in over 80% of attacks. (For the Sparrowhawk the prey was not flying at the start of over 95% of attacks.) Cresswell also noted that the start position of both the raptor and the prey differed with different prey species (Table 14).

Raptor	Prey species	Raptor Position		Prey Position	
		Air	Perched/ Ground	Air	Perched/ Ground
Eurasian Sparrowhawk	Redshank	-	-	13	464
	Dunlin	-	-	6	74
	Skylark	-	-	14	92
Peregrine	Redshank	108	18	14	105
	Dunlin	42	13	35	28
	Skylark	11	8	17	2
Merlin	Redshank	11	27	2	44
	Dunlin	24	90	13	108
	Skylark	30	178	60	172

Table 14 Position of raptor and prey at the start of an attack. From Cresswell (1996) for raptor attacks in east Scotland.

As noted above, Cresswell (1996) found that 50% of Merlin attacks lasted less than one minute, but again the duration of attacks differed between the three species – Table 15 overleaf.

Raptor	Hunt Duration (minutes)				
	<1	**1-2**	**2-3**	**3-4**	**>4**
Eurasian Sparrowhawk	634	100	22	10	1
Peregrine	83	45	15	9	40
Merlin	171	75	44	17	33

Table 15 Duration of hunts for three raptors in east Scotland. From Cresswell (1996).

Though the hunt durations are significantly different, with those of the Sparrowhawk being substantially shorter on average, most attacks for all raptors lasted less than one minute. The longest recorded hunts were those of Merlin on Skylarks. Cresswell considers that use of shorter hunting flights is a means of reducing energy input and the risk of injury during the winter season, and hence improved the chances of survival in order to breed again.

Overall, Cresswell noted that the Merlin success rate was 10.7%, though the individual rates for different prey differed: Merlins were more successful against Dunlin (14.2%) than Skylarks (12.1%) and had no successes in 48 attacks against Redshank. Cresswell also noted that not all attacks in which the Merlin captured the prey led to a kill, 2 of 19 captured Dunlin and 1 of 31 captured Skylarks escaped: both captured Redshanks got away. The Merlin's overall success rate compares well with that of the other two raptors – Eurasian Sparrowhawk 10.7%, Peregrine 8.7% (as might be expected, larger falcons being less able to capture smaller, more agile prey) – but both falcons had a better record of turning captures into kills than the Sparrowhawk which lost 10 of 75 captured Redshanks (though only 1 of 12 captured Dunlin and none of three captured Skylarks).

The success rate noted by Cresswell (1996) compared well with an earlier study at the same site (Cresswell and Whitfield 1994) which noted 8.8% success for Merlins, 11.6% for Eurasian Sparrowhawks and 6.8% for Peregrines averaged across three winters. What was striking about the earlier study was the damage inflicted on the local wader population. In the winter of 1991-1992 Dunlin mortality was 21.2% of the population, Common Snipe mortality was 25.0%, while the Redshank population was reduced by 57.3%. Most of the waders killed were juveniles, Cresswell and Whitfeld suggesting that this was, in part, due to their exclusion from low-risk feeding areas: around 90% of the juvenile Redshank population was killed during the 1991-1992 winter. Cresswell and Whitfield suggest that the high mortality figures were due, in part, to kleptoparasitism (food piracy) of the raptors by Carrion Crows (*Corvus corone*). Overall the crows took 24.5% of all kills: Eurasian Sparrowhawks were the most sorely affected, with Cresswell and Whitfield suggesting that an individual raptor might need to catch 50% more prey than necessary for survival as a consequence of crow piracy.

Once captured, the standard way for falcons to dispatch prey is with a bite to the back of the neck. Limited observation of Merlins suggests that this is the technique employed against smaller prey. However, a famous English book on falconry (Salvin and Brodrick 1855) suggests that for larger birds the Merlin clutches the prey's throat, killing by suffocation. Once dead, prey is taken to a favourite plucking post where, as noted above, the feathers are removed, as are the head, wings and legs/feet. The skin is then pulled back so that the flesh is revealed for consumption. After feeding the bird cleans its bill by wiping it on the perch, then cleans its feet with its bill.

Flight Speed

Bond (1936) noted that a falconry Merlin was twice timed over a distance of 1542ft (470m) and that its speed was calculated at 29.9miles/hr (47.8km/hr). Bond stated that by comparison to prey species Merlins 'flew faster than Quail or Meadowlarks, and more slowly, at least in a rising flight, than Horned [Shore] Larks. It could catch a shrike in a long course free from cover; it was keener after shrikes than any other bird. It could catch, bring down and kill a dove, or even a strong adult common pigeon if released within 50 feet, but was easily outdistanced by these birds after they had attained top speed.' In chasing these prey species Bond guessed the Merlin was about 50% faster than when it flew to the lure, i.e. a speed of about 70km/hr.

The flight speed of Merlins (and Peregrines) was measured by Cochrane and Applegate (1986) during migration flights in the USA by both falcons, using a combination of radiotelemetry and visual observation. Noting that both species soared during migration and rarely flew in straight lines when hunting, but collected data from periods of flapping flight as noted in Table 16 overleaf.

Falcon[1]	Duration[2] (minutes)	Ground Speed (km/hr)	Air Speed (km/hr)
I/m Merlin	20	48.5	48.5
I/m Merlin	28.5	30	30.5
A/f Merlin	30.5	74	44
I/f Merlin	40.5	30	44
	Mean	**45.6**	**39.3**
A/f Peregrine	55[3]	50	49
I/f Peregrine	83	49.5	36.5
I/f Peregrine	9	53	53.5
I/m Peregrine	9	36.5	36.5
I/m Peregrine	17	70	50.5
I/m Peregrine	129.5[3]	35	36
	Mean	**48.8**	**43.7**

Table 16 Mean flight speeds of migrating Merlins and Peregrines measured on straight paths to the nearest 0.5km/hr. From Cochrane and Applegate (1986).

Notes:
1. A=Adult, I=Immature, f=female, m=male.
2. Time over which speed was calculated.
3. Falcon was flying in a strong cross-wind.
4. In a study in Europe the wing-beat frequency of migrating Merlins was measured by Bruderer and Boldt (2001). Only two Merlins were observed

in a study which included a total of ten European/North African falcons, the Merlins being noted as having a mean beat frequency of 6.1Hz, corrected to an equivalent sea-level frequency of 5.9Hz.

Juvenile Prairie Merlin, Alberta, Canada.
John Warden.

By what might best be termed a happy mistake I was able to measure the flight speed of a female Merlin. During photography of a Merlin nest site from a hide the male Merlin perched conveniently close on a post with a Meadow Pipit it had captured. Previous examination had suggested that this, and one or two neighbouring posts, were used for plucking prey. Entranced by the falcon and awaiting the opportunity to film the plucking the author was caught unawares when the male called, then picked up the prey in its bill and half-turned (see photos opposite). Realising too late what was going to happen, there was no time to change the camera speed and the resulting photo has the incoming female as a blur. The female was flying along the wire fence line, and by measuring the fence wire separation close to where the female was photographed, and making assumptions about the length of the bird a speed of 31.7-36.4km/hr was calculated. The speed does not seem high until one remembers that the female is taking prey from the male's bill without either hauling him off the post or knocking him off with her wing: it was, in fact, a beautifully executed, high-speed manoeuvre, and very likely at a speed lower (perhaps much lower) than might be expected in a high-speed chase for prey.

I was also able to measure the speed of Merlins using an 'on-board' unit. The unit was developed after observations of Gyrfalcons hunting on Bylot Island, Arctic Canada (as part of work later reported in Potapov and Sale 2005). I noted that Gyrfalcons were not stooping on prey in straight lines. Discovering that the falcon eye was twin-fovea, and that the more visually acute fovea was set at an angle of about 40° to the main linear axis of the bird (i.e. the axis running through the bill, the centre of the head, back and tail) and knowing that falcon eyes do not move in their sockets by more than 2-5°, I calculated the path a falcon would travel if it maintained the deep fovea on the prey : it was a logarithmic spiral of the form found elsewhere in nature (eg. fossil ammonites and Nautilus spp.). Having worked out the mathematics of the falcon flight path, I was chastened to discover that this had already been achieved, and extremely well-presented, by Tucker (2000). What had not been achieved was complete confirmation that the theoretical flight path was actually been utilised by hunting falcons. To explore the path actually being followed a unit was constructed which attaches to the Marshall backpack which many falconry birds wear as it allows a radio finder unit to be carried (to aid recovery of birds that do not immediately return to their owners at the end of flights).

Opposite
The male Merlin calls (*above, left*), then turns and takes the prey in his bill (*above, right*) so that the female can take it from him (*below*).

With the assistance of others, as noted in the Acknowledgements to this book, the unit was constructed on a purpose-designed printed circuit board (pcb). It comprised gps, a gyroscope and x, y, z accelerometer, and was powered by a small LiPo battery. Data were collected at 1Hz (gps) and 250Hz (other systems) and stored on the bird to avoid the weight of a radio transmitter and the problems caused if line-of-sight contact was lost. In preparing the unit it was clear that both mass and profile were important (**Box 12**). While mass would need to be no more than 10% of body weight, an ethics committee considering university proposals for fitting transmitters etc. on birds suggests a limit of 5%, and so this was set as a target. The first units weighed about 10g, which was fine for flying on Peregrines and other large falconry. A later design reduced the weight significantly (to about 6.5g). The first units were used on Peregrine and Peregrine hybrid falcons, though the later units could also be flown by all the British falcons. Data on the flight paths of the four British falcons (Hobby, Eurasian Kestrel, Merlin and Peregrine) will be published later (Sale *et al.* in prep) but preliminary work on the Merlin is given here.

Box 12
Size and mass were, of course, a critical element of the design. The mass of a unit has a direct effect on the bird as it represents an immediate increase in weight without the advantage of an increase in fitness which might otherwise compensate in the longer term. In studies by the Dutch team at the University of Groningen who trained Kestrels to fly specific routes along corridors, i.e. in still air, (Videler et al. 1988a, 1988b), male and female falcons were loaded with 31g or 61g of lead. Unloaded the falcons flew at speeds close to the V_{mr} (maximum range speed) predicted by theoretical models, but as the payload increased speeds declined, approaching the V_{mp} (minimum power speed) predicted by the same models. As the payload mass increased, the decline in speed was accompanied by an increase in wing beat frequency and an increase in tail inclination. However, the payloads carried by the falcons were large. The male weighed 160g, the female 190g, so 31g represented 19% and 16% of body weight respectively for male and female, while 61g represented 38% and 32%.

In the wild smaller alterations in body mass are likely on a daily basis. Hamby et al. (2004), looking at the variation in flight energy costs and flight speeds in the Cockatiel (*Nymphicus hollandicus* – an endemic Australian species also known as the Quarrion – noted that daily weight changes of 5-10% were normal. Hambly and co-workers noted declines in flight speed of the birds with payloads of 5%, 10% and 15%, but then a reversion to control flight speed (i.e. 0% payload) at 20%. The decreases in speed for lower payloads were initially small (of order 1% at 5% payload, 3% at 10%

Box 12 *continued*
payload), but rose to 7% at 15% payload, a statistically significant difference). The reversion to normal flight speed at the higher payload was accompanied by a significant increase in wing beat frequency: the Cockatiels were changing their behaviour rather than simply reducing their speed. The energy cost of flying increased with payload, but were not significantly higher than unloaded flight for any payload. While this is surprising, Hambly et al. note that Cockatiels regularly increase their body mass by 20% prior to migration, suggesting that the observed behavioural change at the higher payload was an inherent strategy.

While falcons differ from Cockatiels in many ways, the results of Hambly et al. suggest that provided the mass of a unit added to a falcon was minimised the effect on flight characteristics would be limited. This was confirmed by a study by Pennycuick et al. (2012) using Rose-coloured Starlings (*Sternus roseus*), though a separate issue was noted in that work, namely interference with the air flowing over the back of the bird. In the work of Pennycuick et al. the payload (a dummy transmitter) projected 6mm above the back of the bird. This increased the drag coefficient of the bird (as measured in a wind tunnel) by 50%, but the addition of an angled aerial increased the drag coefficient by almost 200%. Pennycuick and co-workers then simulated the effect of the mass and drag coefficient increases by considering the actual flight of a Barnacle Goose (*Branta leucopsis*) which had been fitted with a satellite transmitter to study its migration. The simulation suggested that the effect of increased mass due to transmitter would have been small, but the effect of drag coefficient increase would have reduced range and decreased energy reserves on arrival.

A related concern was raised by Sodhi et al. (1991b) who wondered if the radiotagging of urban-breeding Merlins in Saskatoon was affecting breeding success or survival, particularly as other studies (on game birds) had noted reductions in both, as well as abnormal behaviour in tagged birds. The tags used weighed 4g, representing 2.4% and 1.6% of the body weight of male and female Merlins respectively. In a study in which the breeding success of tagged and untagged males was measured, there was no difference between the two groups. In addition, survival rates of both males and females, as measured by the return rates of the birds the following year, were indistinguishable. By observation there was also no discernible behavioural difference between tagged and untagged birds, though those fitted with tags (either to the underside of tail feathers or legs) tended to peck at the tags for a period and to preen frequently immediately after release. In conclusion Sodhi et al. conclude that the small radiotags were having no effect on the Merlins.

The unit was flown on a male Merlin called 'Atom' (because he was tiny, but feisty, a 'mighty atom'). Fig. 3 shows a typical flight. Ideally Atom would have found a lark and the pair would have engaged in a ringing flight, but with few falconry Merlins available in the UK such luck would have been outrageous even though the habitat (in southern England) was ideal for larks and there were many about during the illustrated flight. In that flight there is evidence to suggest that Atom engaged in a short chase of larks (at Points A and B), and rather less conclusive evidence at Point C (all these points were out of view of the falconer and myself as they occurred beyond a tree belt in an adjacent valley). During the flight the wind was steady at 5.5m/s, but occasionally gusted to 7m/s. A correction to windspeed is required to allow for the increase with height above the ground (due to frictional effects). With that correction included the maximum air speed reached by Atom during the flight was about 20m/s (a little over 70km/hr) during a short stoop: in level flight Atom rarely achieved more than 10m/s (about 35km/hr). Both speeds are comparable to those observed or suggested above, but the proviso, as always with falconry birds, is that the level of fitness of 'domestic', as opposed to wild, falcons may not be comparable.

Male Merlin 'Atom' after returning to the lure at the end of a flight. The flight data unit can be seen behind his head.

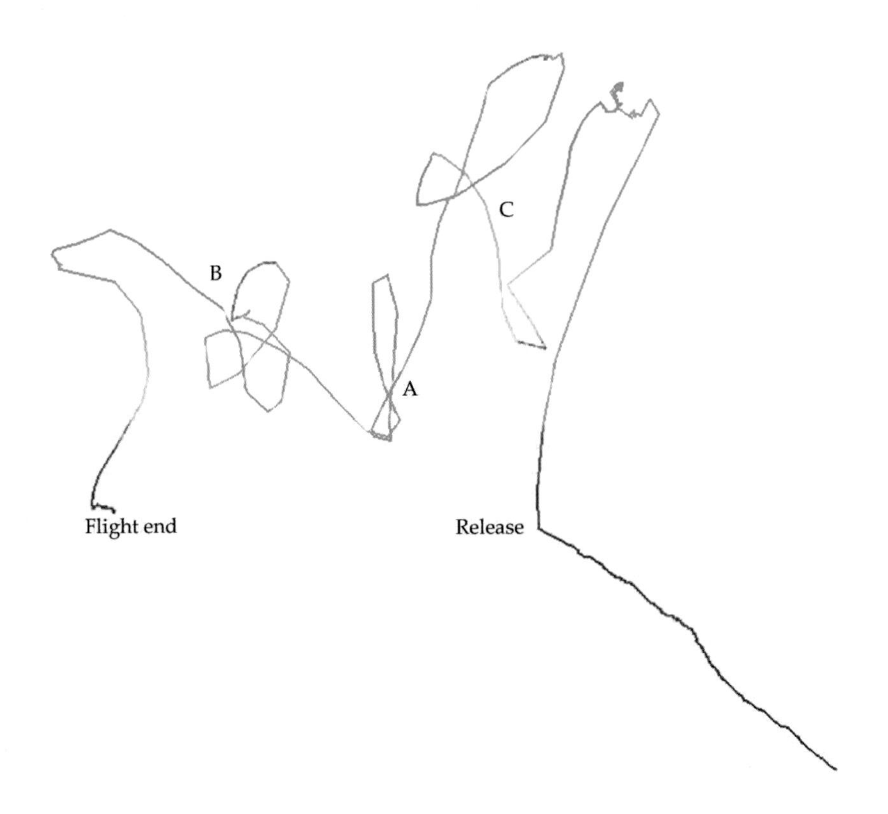

Figure 3
Plan (*above*) and profile (*below*) of the Merlin flight considered in the text. The variation of colour along the flight path allows the two images to be matched.

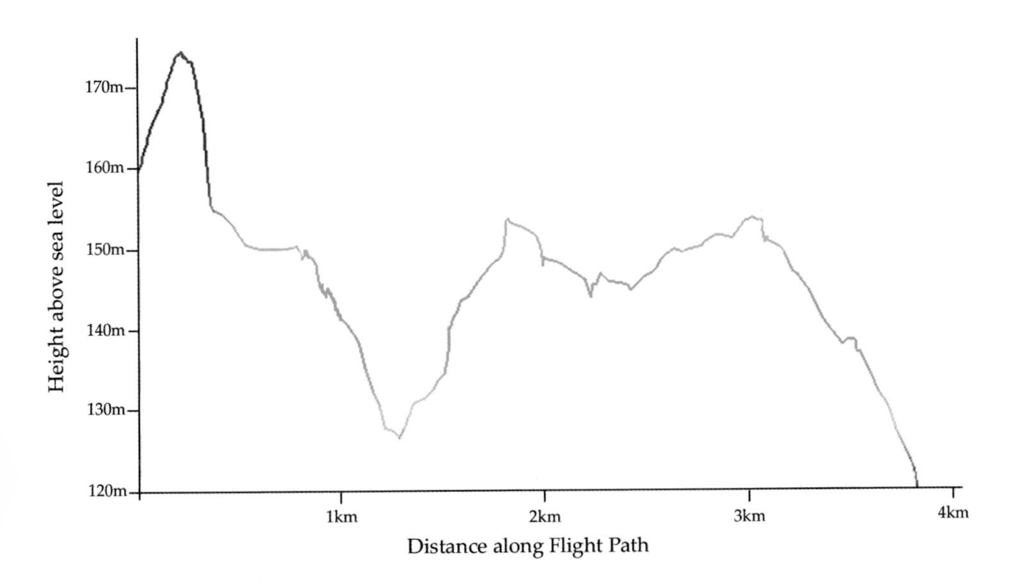

Prey Defence

The success rates noted earlier in the Chapter imply good defence techniques by potential prey. One such defence is flocking. Buchanan *et al.* (1988) noted three flocking defence techniques of Dunlin. In flashing flight the Dunlin changed direction or tilted their bodies synchronously so that the dark upper and pale lower side were flashed successively. In rippling flight the body movements were not synchronous, a wave of movement passing through the flock. Finally, in columnar flight the flock rose in a 'tornado-like vertical column'. Combinations of the techniques also occurred: in all cases both flashing and rippling might be observed. The various techniques presumably confuse the predator, and each seems successful: flashing which was used in 64% of attacks had an 85% success rate at avoiding a Merlin attack; flashing and rippling occurred in 28% of attacks and had a 91% success rate; while a combination of columnar flight, flashing and rippling occurred in 8% of attacks and had a success rate of 71%. Boyce (1985), in a winter study at Humboldt Bay, California, added a further dimension to anti-predator tactics of flocks. He noted that often a Merlin would knock a prey individual (Dunlin was again the main target) into the water rather than capturing it or, occasionally, the Dunlin would deliberately dive into the water to escape capture. If the Dunlin survived being hit and could still dive as the Merlin came around in an effort to catch it the wader might survive. (Brady (1999) noted a Spotted Sandpiper (*Actitis macularia*) performing the same escape manoeuvre in the Chequamegon Bay inlet of Lake Superior). On the occasions Boyce (1985) saw a Dunlin dive to escape capture, and as the Merlin slowed in order to return to seize the Dunlin from the surface, the flock wheeled around and positioned itself over the water-bound bird, shielding it from the Merlin and giving it a chance to return to the flock. Often this tactic succeeded, though if the downed bird was injured it obviously did not. (As a digression, Duncan (1990) reports a Merlin capturing a Eurasian Dipper (*Cinclus cinclus*) 30m from shore above a Scottish loch, then (accidentally) dropping the prey and circling to retrieve it from the water surface).

If flocks are such a successful anti-predation strategy, then it would be anticipated that Merlin attacks on flocking birds would be limited, but in his Scottish studies Cresswell (1996) found that all three raptors attacked larger flocks of Dunlin more frequently than would have been expected by chance occurrence (Fig. 4). This suggests that although a large flock enhances the chances of a prey species individual surviving a raptor attack, they also improve the raptor's chances of a kill. But Cresswell's data did not support the idea that the raptors chances of a kill were improved, as for both Merlins and Sparrowhawks the attack success rate was better against small flocks (defined as 1-10 birds) than against larger

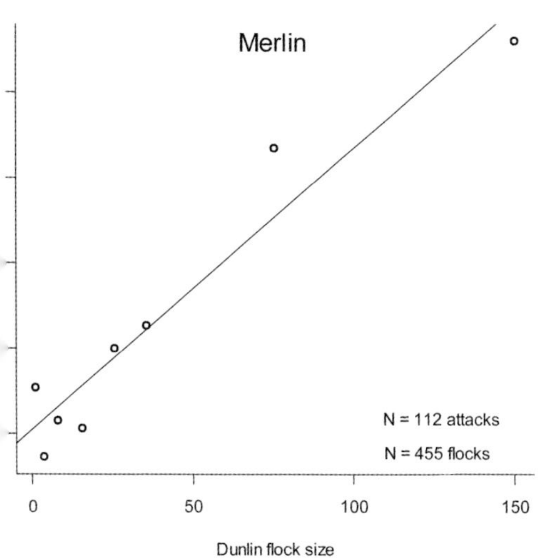

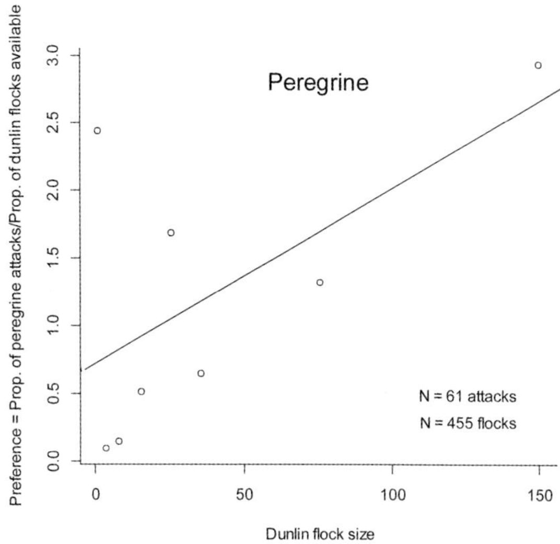

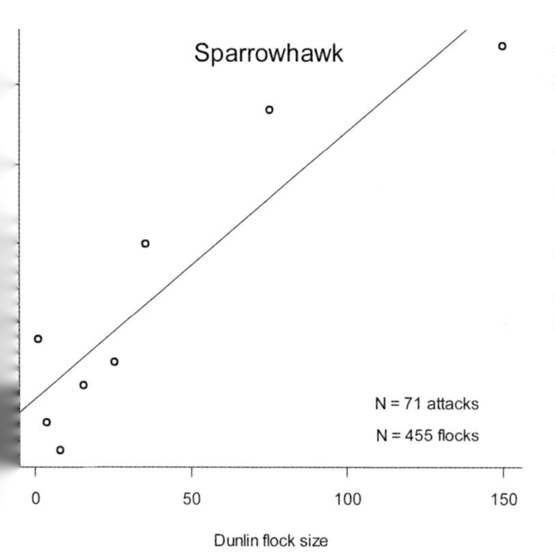

Figure 4
Attack preferences for different flock sizes of Dunlin being hunted by Merlin, Peregrine and Eurasian Sparrowhawk in East Lothian, Scotland. A preference of less than 1 indicates avoidance; 1 indicates flocks were attacked according to their availability; greater than 1 indicates flocks were preferentially attacked. Redrawn from original data supplied by Will Cresswell. For further details see Cresswell (1996).

ones (defined as 11-200 birds): for Peregrines the success rate was similar. However, the difference in capture rates for the two smaller raptors between small and large flocks was not significant, so the result is not counter-intuitive (i.e. it does not necessarily imply that raptors appear to consider more targets improves the likelihood of a kill even though defence strategies by flocks are successful, an idea which would imply that raptor attack strategies are not influenced by experience). The more likely explanations are that raptors attack larger flocks as this increases their chances of spotting a weak bird, and that flocking as a defence has its limitations, particularly for species such as Dunlin which are less successful at maintaining close separation and the cooperation needed for complete safety. Invariably some birds are left at the periphery of the flock where they are vulnerable to attack (**Box 13**). For other species flocking is an improved defence technique (**Box 14**).

Attacking flocks may have its attractions for the falcons, both Peregrines and Merlins, but it also has drawbacks. Buchanan (1996) observed three occasions for each species when the falcon hit the water before pulling out of a stoop: he does not say whether the falcons survived, though the implication of there being no mention of the outcome is that all six birds did.

Box 13
Attacks on the main body of a flock can also be successful. McIlwraith (1886) records an incident observed in Ontario, Canada in delightful prose, noting that 'in the fall, when the Blackbirds get together in flocks, they are frequently followed by the Little Corporal who takes his tribute without much ceremony' and that the Merlin – which McIlwraith usually calls the Pigeon Hawk: it is not clear why on this occasion he chooses to salute the bird as the new Napoleon – dived into the flock and 'came out on the other side with one in each fist'. McIlwraith is not specific as to which 'blackbirds' he is referring, but from his descriptions of other species it would have been either the Red-winged Blackbird or the Rusty Blackbird (*Euphagus carolinus*).

Flocking is one form of defence against predation, but prey species have other alternatives. One obvious one is that they can seek cover, either by crouching on the ground or, for waders, diving. Cresswell (1993) noted these strategies, as well as prey outflying the predator, by observing the escape responses of Redshanks under attack by Merlin, Peregrines and Eurasian Sparrowhawks in eastern Scotland. Cresswell noted that 22% of Redshanks which crouched were captured by attacking Merlins, while all waders which flew or dived escaped capture. For the other two predators the figures were 8% capture if flying, 20% if diving and 91% if crouching (Eurasian Sparrowhawk), and 14% flying, 2% crouch or dive for Peregrines. However, it should be noted that very many more Sparrowhawk attacks than either Peregrine or Merlin attacks were observed – for instance over 500 Redshank responses ot Sparrowhawk attacks were observed, compared to about 150 for Peregrines, but only about 50 for Merlins. What was clear however, despite the difference in

Box 14
Starlings are renowned for their flocking abilities, the black clouds of the birds moving synchronously against the night sky as they come in to evening roosts being one of the magical sights for the birdwatcher. That the seemingly choreographed movements are a defence strategy has long been assumed, but in an interesting study in Rome Carere *et al.* (2009) were able to show that not only was this the case, but that there was a social behavioural side to flocking. Two Starling flock areas were studied, a city centre square with rows of trees surrounded by buildings where about 20,000 birds collected, and a more open parkland area south of the city where 50,000 birds gathered. The predation risk (from Peregrines) was lower in the city centre area than in the parkland. Carere and co-workers catalogued a series of flock forms, and found that larger, more compact forms were found in the high predation risk area, whereas smaller, looser flocks and singleton birds were found in the low risk area. Peregrine success rates in the low risk area were higher. While these findings are intuitively unsurprising, what was interesting was that the Italian researchers found that the behaviour of flocks at a distance from the roosts influenced local behaviour, i.e. if a distant flock showed anti-predator flock forms, the local birds adopted similar forms, implying that social information was being passed between flocks.

numbers of observed attacks by the three predators, was that the Redshanks were altering their escape behaviour dependent on which predator attacked: the risk of capture in crouching when under attack by Sparrowhawks meant that 86% of attacked Redshanks flew, only 4% crouching. For Peregrine attacks the higher risk of aerial capture meant that over 60% of Redshanks chose to crouch or dive. Similarly, most Redshanks attacked by Merlins flew rather than crouching.

'Atom' enjoying a meal after another successful flight.

Another escape strategy noted by Cresswell in a separate study (Cresswell 1994) was the use of song as an indication to the predator that the individual prey bird is in good condition (**Box 15**) in the hope that the predator will become disheartened and give up the pursuit. Cresswell studied the interaction of Merlins and Skylarks during winters on the Tyninghame estuary in East Lothian, Scotland. When attacked a Skylark may not sing, sing poorly or sing well. Cresswell's data shows that both the Merlin's chase time of a prey bird and the capture rate increased significantly depending on the Skylark's singing behaviour (Fig. 5). Note that although Fig. 5a implies that Merlins preferentially attack birds which can sing, either well or poorly, the Skylark will not start to sing (if it can) until it is attacked, so the Merlin has no *a priori* knowledge on the singing abilities of the bird until it attacks it as Skylarks are normally silent in winter: the relative frequencies of attack are therefore more a measure of the overall condition of local Skylarks than the Merlin's ability to choose its victim, particularly as Cresswell found no apparent difference in the flocking ability of Skylarks based on their ability to sing. That singing is an anti-predation strategy was apparent not only from the statistics of Merlin attacks (i.e. that they were less likely to pursue a singing Skylark) but from the deployment of singing against differing predators. Skylarks do not sing when attacked by (Eurasian) Sparrowhawks. The

Box 15
The idea of pursuit-deterrent calls in avian prey derives from the existence of 'stotting' in certain mammalian prey species. Stotting involves leaping into the air with all four legs held straight and is best seen in the African species Springbok (*Antidorcas marsupialis*) and Thomson's Gazelle (*Eudorcas thomsonii*), and in the Pronghorn (*Antilocapra americana*) in North America. Several theories have been put forward for this curious behaviour including the idea that it is a visual alarm signal to conspecifics or a method by which a group of conspecifics can confuse a predator. It has also been suggested that it alerts a predator that it has been seen and so spoils the element of surprise the predator is banking on achieving. Stotting may occur during a chase which seems to invalidate the latter hypothesis, and also leads to the most interesting theory. This suggests that the stott is a signal to the predator that the prey is in good condition and therefore likely to escape pursuit and so is not worth chasing. Since stotting may occur during a chase it is an 'honest signal' in that it shows a willingness to engage in useless behaviour which enhances the likelihood of capture and as such indicates that the prey is so confident of its condition that it can afford the risk. As birds do not have the capacity to stott, pursuit-deterrent calling may perform the same function. For further details of this idea see Cresswell (1994) and references therein, including Woodland *et al.* (1980) which laid the foundation for the idea in avian prey.

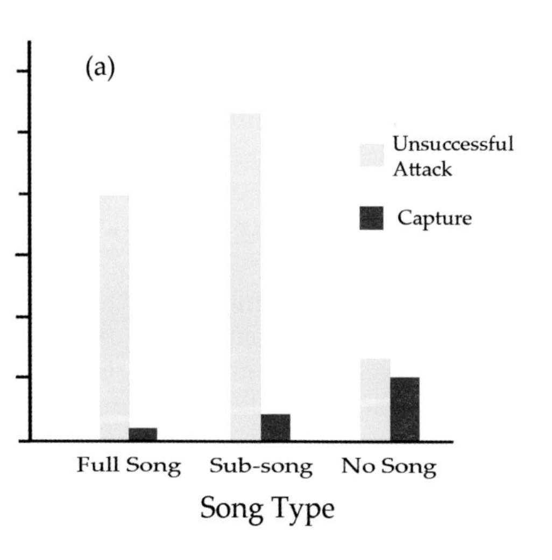

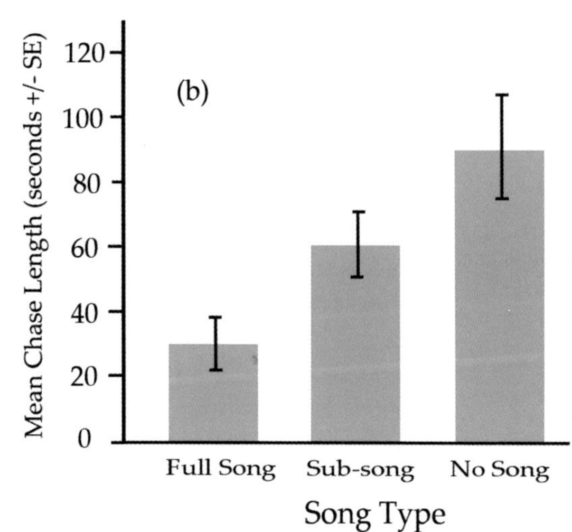

Figure 5

a) Frequency of capture of Skylarks with differing song types.

b) Length of chase time by Merlins of Skylarks with different song types.

Redrawn from Cresswell (1994).

Sparrowhawks have broader, shorter wings, a shape which is most efficient at lower speeds. In contrast, the Merlin wing is better adapted to both faster and more sustained flight. If chased by a Sparrowhawk, the Skylark's best defence is therefore to save its energy and to outfly the raptor, but when pursued by a Merlin, it is better to try to persuade the raptor to give up.

If the Merlin attacked a flock, several birds might sing, but by that time the Merlin would have chosen its intended victim, usually a bird, or one of several, marginally detached from the flock. Once attacked the Skylark would start to sing (if it was going to sing) when the Merlin was within about 10m. While singing is not overly energy intensive, singing while being pursued will inevitably result in either the bird slowing or in becoming exhausted sooner, either of which enhances the likelihood of capture and so is a dangerous strategy. A bird which can therefore sing, and especially one which can sing a full territory-claim song, is clearly signalling its condition – I have energy to spare, chase someone else: Cresswell's data show that Merlins respond by calling off the chase of full singing birds early.

Concerned that other variables might be affecting the Skylarks decision to sing or not, Cresswell looked at the effect of ambient temperature on

Skylark response to Merlin attack. He found that in November, a mild month, 20% of Skylarks were caught on the ground, but that this rate had risen to 60% in the colder months of December and January. As lower temperatures were likely to affect the condition of the Skylarks, this marked difference implies that singing is indeed a defence strategy as any bird which could sing would prefer to risk the chase than to rely on cryptic plumage. A rather more effective escape strategy was observed by Boyle (1991) who had a Skylark dive under his car, brushing his leg as it passed, to escape a pursuing Merlin.

Cresswell (1996) in his study of Merlin, Peregrine and (Eurasian) Sparrowhawk attacks in east Scotland also noted the behaviour of the three raptors in attempts to capture prey by surprise. If an attack fails, the raptors have two choices as a consequence of an initial, failed, attempt having alerted the local prey to danger: either moving to another area or waiting until the vigilance of the prey had declined to a point where surprise became a possibility again. Cresswell found that in general, within 3-5 minutes Redshank had relaxed to the point where a renewed attack would have the same element of surprise as an initial attack. In addition, because Merlin (and Sparrowhawk) are relatively small raptors with limited hunting ranges (in comparison to Peregrines which flew greater distances in search of prey) they needed to move only about 100-200m in order to arrive in an area where they could mount a surprise attack on a different prey population, one not spooked into extra vigilance by an earlier attack. The Merlin could therefore choose to wait a short time, or move a short distance. However, either strategy only works if there is no other Merlin in the area, and Cresswell noted during his study that on 97% of the occasions when he saw two Merlins, one was chasing the other, apparently confirming the idea that to maintain surprise a Merlin must attempt to preserve sole hunting rights to a wintering area.

However, this suggestion that Merlins seek to maintain sole hunting rights does not operate at all times, as Bengtson (1975) on open-country Merlins, and Sodhi and Oliphant (2005) for urban-dwelling Merlins note. Bengtson observed that in 30% of attacks two Merlins (one male and one female in all cases where sex could be identified in both birds) hunted co-operatively. Bengtson notes both surprise and 'persistent chasing' attacks being carried out by pairs, but unfortunately does not define 'persistent chasing': presumably this would involve two Merlins chasing the same prey, the latter's manoeuvres being limited allowing one or other falcon the opportunity of a stoop attack. For the co-operative urban-dwelling Merlins, Sodhi and Oliphant (2005) noted one falcon flying along a tree corridor below the canopy flushing waxwings (the reference is not clear which of the two waxwing species seen in North America was involved, but as it was a winter observation it was probably the Bohemian Waxwing)

Male Icelandic Merlin.
Jóhann Óli Hilmarsson.

which the second Merlin, following behind, then attacked. While Bengtson's (1975) observation was in summer, and so almost certainly involved a breeding pair of Merlin, the observation of Sodhi and Oliphant (2005) was of wintering Merlins. However, Buchanan (2010) records simultaneous hunting by two Merlins but is unclear whether this was actually co-operative. In a total of 253 observed hunts on three occasions (i.e. six hunting flights, 2.4% of the total) he observed two Merlins hunting simultaneously (in one case a female *F.c. suckleyi* and a male *F.c. columbarius*). In these cases Buchanan was unclear whether the hunting was genuinely co-operative as it could also be explained by the second falcon using the confusion in a flock of Dunlin (the intended prey in these attacks) to make its own surprise attack, or could have intended to pirate the first bird if it was successful. Buchanan concludes by suggesting that it is difficult to separate co-operation, fortuitous interaction and kleptoparasitism when observing simultaneous hunting. Dickson (1998c) also reports simultaneous hunting between two Merlins and a Hen (Northern) Harrier, but again it is not clear if this was co-operative or one raptor using the confusion caused by another. Dickson terms the behaviour 'association' which seems a more accurate description than true co-operation.

Food Consumption and Energy Balance

In a fine piece of experimental and observational work Warkentin and West (1990) investigated the energetics of Merlins during the harsh winter experienced in Saskatoon where the falcons are resident. For the work, six female and three male free-living Merlins were captured and radio-tagged. The birds were then observed from the time of leaving their night-time roost sites until their return to them to establish periods of perched behaviour, with the time spent flying being estimated by distance travelled (as recorded by the radio tag transmitters) assuming the birds flew at an average straight-line speed of 50km/h. The daily time budget was assumed to be divided into five activities – night-time roosting which was assumed to cover all the time spent at the roost; daytime perching, which was assumed to be an active state, the bird alert for the possibility of prey capture; preening; eating; and flight. Environmental data (temperature and wind speed) was also collected at the birds' roosts and the degree of vegetative coverage at the roost sites was estimated.

A further nine captive Merlins (five males and four females: one of the males was also part of the radio-tagged contingent) were used to provide metabolic data by testing in a chamber held at a constant temperature through which air at a known rate could be drawn, the air being passed through a desiccator, and CO_2 and O_2 analysers so that metabolic heat production could be calculated for various states of activity (i.e. resting, eating and preening).

To compute the energy intake of the Merlins in the study, the common prey taken by the free-living falcons (House Sparrows and Bohemian Waxwings) were trapped, processed (the head, legs/feet and wings removed, the main feathers discarded, as in the standard form of pre-consumption processing carried out by the Merlins) and then analysed (in a calorimeter) to determine caloric value. The same process was used on Japanese Quail which were fed to the captive falcons.

In carrying out a similar experiment to that of the Canadians in order to establish the total energy expenditure of the Eurasian Kestrel, a Dutch team at the University of Groningen (Masman *et al.* 1988a) added the components of the equation:

$$E = B + T + A + H + S \text{ (kJ/day)}$$

Juvenile Prairie Merlin, Alberta, Canada.
John Warden.

where B is a basal component, the energy required to keep a thermoneutral, fasting bird alive;
T is the energy required for thermoregulation (i.e. to overcome heat loss). The Dutch actually calculated not only heat loss for a fully-fledged bird, but also the additional heat loss by a bird in moult (Tr) as feather loss increases heat loss;
A is the energy required for activity, which was sub-divided as Ab (activities other than flying), Af (flight) and Ah (flight-hunting);
H is the energy required for feeding (i.e. digestion);
and S is the energy required for tissue synthesis (eg. feather synthesis during moulting).

Warkentin and West (1990) used a modification of this equation, but ignored tissue synthesis as the wintering Merlins in their study were not moulting and other synthesis was considered to be negligible. As Merlins do not 'flight hunt' (flight hunting is the familiar hovering of Kestrels, a hunting technique not shared by Merlins) the Canadian authors did not need to split 'A' into sub-sets. However, in addition to the experiments to establish the Merlin's time energy budget (TEB), Warkentin and West also used environmental data to investigate the Merlin's thermal neutral zone (TNZ): in particular the Canadians were keen to establish the lower critical temperature of the TNZ, the temperature below which the bird does not have to increase its metabolic rate in order to compensate for a reduction in ambient temperature.

The time budget of the free-living Merlins is set down in Table 17:

	Resting at Perch Mean (range)	Alert at Perch Mean (range)	Eating Mean (range)	Preening Mean (range)	Fligh[illegible] Mea[illegible] (rang[illegible]
Female Merlins					
a) as % of 24hr day	60.5 (51.6-65.6)	34.2 (32.9-35.6)	3.2 (1.5-4.0)	1.3 (1.0-1.9)	0.9 (0.6-1[illegible]
b) as % of active phases	0.0	86.5 (83.3-90.0)	7.9 (3.7-10.2)	3.3 (2.6-4.8)	2.3 (1.4-3[illegible]
Male Merlins					
a) as % of 24hr day	59.6 (54.6-64.5)	34.1 (31.2-38.3)	2.4 (0.6-4.8)	2.7 (1.1-3.0)	1.2 (0.4-1[illegible]
b) as % of active phases	0.0	84.4 (77.3-94.7)	5.9 (1.6-12.0)	6.6 (2.6-7.5)	3.1 (1.1-4[illegible]

Table 17 Time budget of wintering Merlins in Saskatoon. From Warkentin and West (1990).

Notes:
1. Figures in parentheses below each mean value are the range of observed percentages. Ranges are larger for male birds because of the smaller sample size.

2. The percentages of activity phase are very similar for male and female Merlins, and are not statistically different, apart from the larger time fraction spent in body care by male Merlins. This could imply that male birds are more fastidious, but is probably more likely to arise from the small male sample group (and see the note to Table 18 overleaf).

Warkentin and West found that the TNZ of female Merlins was wide, extending from 3.4°C to at least 20°C, and that within the TNZ the basal metabolic rate was 6.96±0.23kJ/hr (equivalent to a mass-specific rate of 7.92±0.20mW/g, the average mass of the tested birds being 241g). The male Merlin basal metabolic rate was lower at 5.23±0.19kJ/hr (equivalent to a mass-specific rate of 8.49±0.28mW/g, the average mass of the tested birds being 176g), but the male TNZ was much more limited, the lower critical temperature being 14.3°C (with, again, an upper limit of at least 20°C). The lower critical temperature, and its accompanying increase in metabolic rate, means that roosting Merlins seek to limit heat loss during the night and Warkentin and West noted that the birds, when seeking a roost position, often took care to position themselves on the leeward side of their chosen tree to reduce convective heat loss and also chose positions with vegetation coverage. The difference, in terms of total energy expenditure, between a protected and an exposed roost was found to be about 6%, which is the equivalent of about 22km of flight, explaining why some birds were seen to fly up to at least 15km from preferred daytime rural hunting areas to roost in the city. While choosing a leeward roost with good cover will aid the reduction in convective heat loss, it could also be a survival strategy. In Europe Merlins are known to be victims of owl attacks (Mikkola 1983 who notes both Eagle and Tawny Owls (*Strix aluco*) killing Merlins) and it can be assumed will also be taken by large owls in North America: it is certainly the case (Hodson 1976) that Merlins, both adults and nestlings, are predated by Great Horned Owl (*Bubo virginianus*).

The difference in lower critical temperature between the Merlin sexes is extremely interesting, as Warkentin and West note, as it implies that males and females have chosen different physiological routes in order to combat the cold of winter (and the cold of their preferred, northerly, habitats during all periods of the year). While females have increased their TNZ, males have not, instead reducing their thermal conductance (i.e. by increasing the insulating properties of their plumage) and so reducing heat loss.

Warkentin and West also calculated the variation of energy costs of the various phases of Merlin activity (Table 18 overleaf).

Activity	Female Merlins		Male Merlins	
	Energy Cost (kJ/hr)	**Cost as multiple of Basal Rate**	**Energy Cost (kJ/hr)**	**Cost as multiple of Basal Rate**
Basal Rate	6.96±0.23	1.0	5.23±0.19	1.0
Alert Perch	10.05±0.43	1.4	10.13±0.31	1.9
Eating	19.34±1.99	3.3	19.67±1.31	4.3
Preening	22.65±1.48	2.8	22.47±1.02	3.8
Flight	51.47	7.4	33.25	6.4

Table 18 Energy costs of activity phases as a multiple of Basal Metabolic Rate. From Warkentin and West (1990).

Notes:
1. There is no statistically significant difference in the multiples between males and females, implying that male Merlins are not more fastidious than females. As can also be seen, flying is hard work.
2. The overall energy expenditure for a female Merlin was calculated as 305.6kJ/day, and as 279.3kJ/day for a male. The figure for the females equated extremely well with the daily energy consumption for the female Merlin, as calculated from the biomass of consumed food and its energy content. While the figure for male Merlins equated less well, it was in reasonable agreement. What is surprising is that the two figures are so similar given the difference in body mass between male and female. This finding is very similar to that for Eurasian Kestrels in the work of the Dutch group referenced above. In both Eurasian Kestrels and Merlins it seems that males have lower basal energy costs, but higher thermoregulatory costs, the combination meaning their overall costs are similar to those of females.
3. Merlins, it would appear, do not consume prey in order to increase bodily fat reserves, presumably because this would reduce the efficiency of hunting. In their study of Eurasian Kestrels, Masman *et al.* (1986) noted that the falcons were very dependent on cached food when weather conditions did not allow hunting. It is probable that Merlins do the same.

Finally, before leaving the work of Warkentin and West (1990), it is worth noting some interesting information provided by the authors. Considering the energy requirements of Saskatoon's Merlin population (38 birds at the time of the experiments/observations) the authors estimate that the falcons would have consumed 7,524 House Sparrows (or the equivalent of that number in sparrows and waxwings) during the 120 day winter period. This equates to 1.65 sparrows/falcon/day. Assuming the 38 falcons were 19 males and 19 females, then this suggests each falcon is consuming about 25% of its body weight per day. Page and Whiteacre (1975) in their work on wintering Merlins found that the female they were observing was taking prey which weighed an average of 71g each day. Assuming an average weight female, this equates to 28% of body weight daily, in good agreement with the Saskatoon figure. Using such figures, and corresponding values for the feeding requirements of a brood of four chicks suggests that the falcon family would consume around 1000 30g birds during the 100 day breeding season.

Warkentin and West (1990) also considered whether the stresses placed on the Merlins by the Saskatoon winter were close to the limit which the birds could survive. To do this they measured the activities of a female Merlin on days when ambient temperatures were -17°C to -30°C. On these days the falcon ate four sparrows, taking about 30 minutes to ingest each and 45-60 minutes to empty her crop. Given that daylight lasted about 9 hours, allowing for flight time it seems that on such days the falcon was approaching the maximum energy intake in the time available: the Saskatoon Merlins might be living at the limit of the winter range in terms of ambient temperature and may only be able to survive because of the abundance of prey. Observation of the female Merlin also suggested that the very large percentage of the bird's time budget apparently spent idling may be nothing of the kind, the bird being actively engaged in digestion, though this is not to say that Merlins (and other species) in winter do not attempt to minimise their energy output by avoiding unnecessary activity. In doing so Merlins would be reflecting another finding of the Dutch (Groningen University) group that studied the Eurasian Kestrel who found (Masman *et al.* 1988b) that in summer the Kestrels maximised their daily energy gain in order to maximise their reproductive output, while in winter they minimised their energy expenditure in order to minimise their daily energy requirements.

As we have already seen from the work of Warkentin and West (1990) considered above, male Merlins spend only 1.2% of the day in flight, with females spending even less, at 0.9% – see Table 17. Migrating birds will, of course, spend a greater percentage of their day in flight, but during stopovers when they are refuelling prior to continuing the long journey north they may also spend longer in the air as they need to increase their energy intake. In their study of a female Merlin breaking her migration north for a seven-day stopover on the island of Loggerhead Key in the Gulf of Mexico Raim *et al.* (1989) noted that the falcon, a female, which had been trapped and radio-tagged, made an average of 39.6 flights each day, totalling a mean of 36.7 minutes, 5.7% of the daylight hours (Fig. 6).

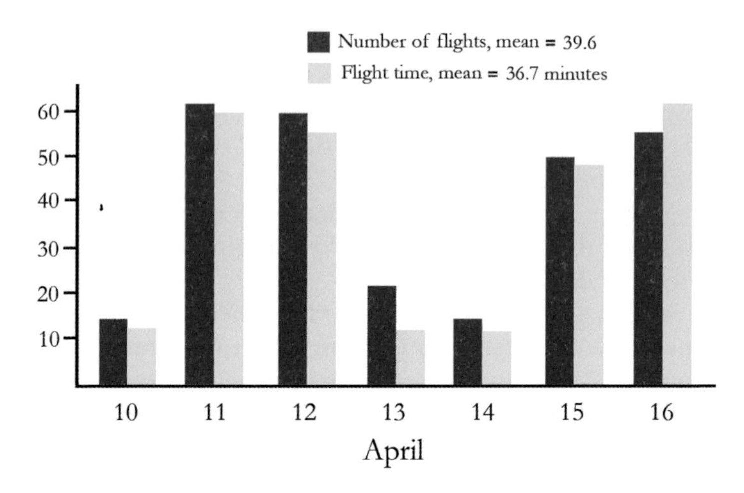

Figure 6
Number of flights and total flight times of a female Merlin making an island stopover on Loggerhead Key, Gulf of Mexico.

Notes:
1. The number of flights and the total flight times are extremely similar: on average attack flights (both failed and successful) and flights between hunting areas average 60 seconds.
2. In addition to spending 5.7% of daylight hours flying, the Merlin also spent 7% of her time eating. With smaller prey this comprised time preparing, eating and bill cleaning and if only smaller birds were captured the eating time averaged 4% of daylight hours. With larger prey, such as Yellow-billed Cuckoo, the eating time percentage rose as high as 14.6% as the Merlin would stand with the prey between bouts of eating.
3. The Merlin also spent an average of 3% of daylight hours preening (range 0.14-0.8%). Including in this time were attempts spent at removing the radio-tag, though these were short (and unsuccessful).
Redrawn from Raim *et al.* (1989).

During the breeding season the time spent by male Merlins in flying rises considerably as the falcon must provision both himself and his mate and, later, provide food for growing nestlings. The male delivers a single prey item on each occasion he visits the nest and can either deliver it whole, or prepared (i.e. plucked with head, legs/feet, and wings removed). In a study of Merlins near Saskatoon, Sodhi (1992) used data from a radio-tagged male Merlin to investigate the preferred option with a diet which comprised mainly House Sparrows. The mean weight of a House Sparrow is 28.6g, of which a Merlin preparing prey will discard 5.7g (20% of total mass). Sodhi found that in general prey that was caught at a distance from the nest was prepared, while that caught closer to the nest was more likely to be delivered whole (Fig. 7). While this result is as might be expected considering the energy cost of flying, Sodhi calculated that the male would reduce its daily flight costs by only about 3% by adopting the strategy, which perhaps explains why Fig. 7 shows a less categorical difference than intuition might suggest. (However, while small, such energy savings may be significant over the entire breeding season when the male has to do all of the hunting for the pair for perhaps eight weeks.) In addition, Sodhi noted that the energy cost of preparing the prey was

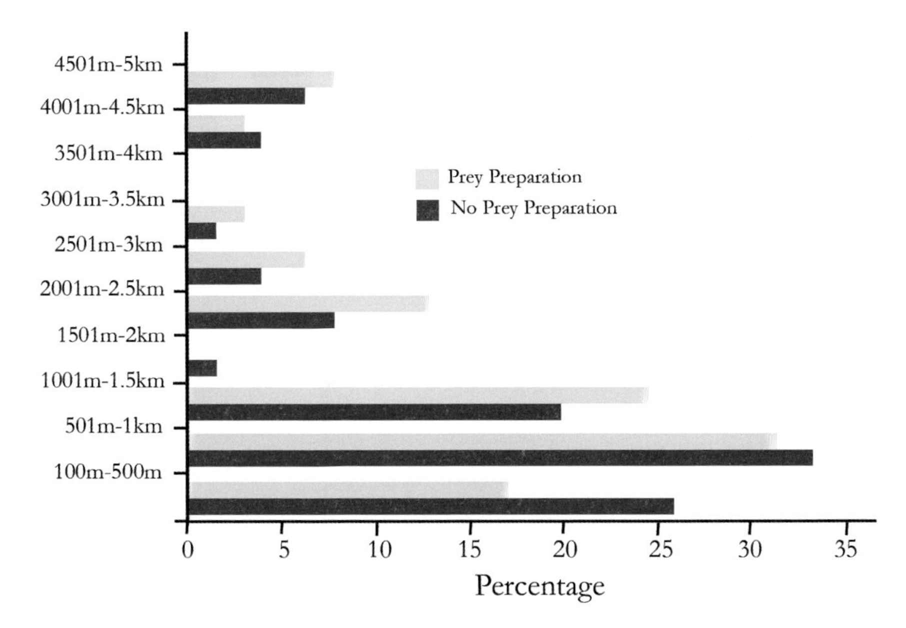

Figure 7
Percentage of times prepared and unprepared prey was delivered to the nest by male Merlins as a function of distance of prey capture from the nest.
Redrawn from Sodhi (1992).

about 50% of that required for unloaded flying (over the same period of time), so by delivering a greater fraction of unprepared food when he makes a kill close to the nest, the male makes an energy saving as the prey is then prepared by the female (both for herself and the nestlings), the trade-off being the increased energy cost of flying with the entire prey against the energy cost of preparation.

But despite choosing a delivery strategy which reduces the energy costs of hunting, the male Merlin spends much more time flying. Sodhi (1993a) trapped male Merlins in Saskatoon, Canada and fitted them with radio tags which allowed the frequency and duration of flights to be recorded. The number of flights (which were assumed to be hunting trips), the duration of flights and the percentage of daylight hours spent flying during the three phases of breeding (incubation, nestling and fledging) are shown in Fig. 8. As the breeding season advances, the male's duty

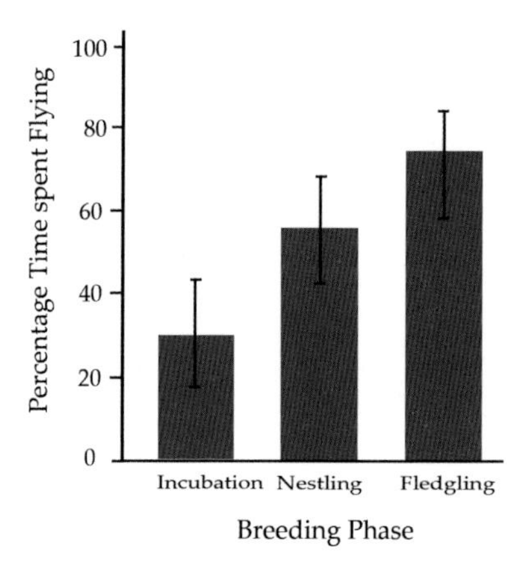

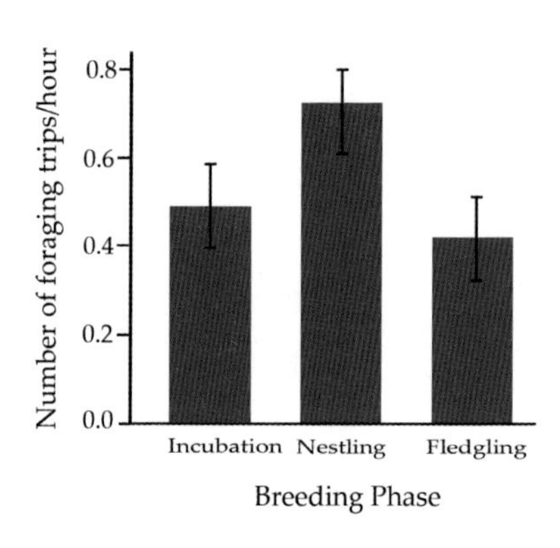

Figure 8
Hunting flights by male Merlins through the phases of the breeding season. Each bar of the histograms represents the mean for all males observed during each phase, together with standard errors.
Redrawn from Sodhi (1993).

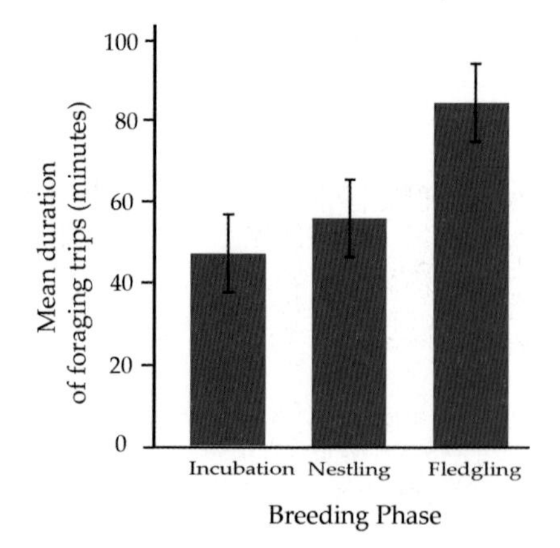

increases, as seen by the increase in the percentage of time spent flying. What is interesting from Fig. 8 is that as the duration of hunting trips increases through the phases of breeding, the number of hunting trips rises from the incubation to the nestling phase, but then falls – longer hunting flights mean fewer can be made. Sodhi does not quote the actual values of the percentages spent flying in the three breeding phases, but does give them in Sodhi (2005) where they are 30.9±2.3% of daylight hours during the incubation phase of breeding, this figure rising to 57.5±4.3% during the nestling phase, and to 70.8±3.9% during the fledgling phase. Sodhi (1993a) also notes that his radio-tagged birds showed, as would be expected, that those males with prey-rich hunting ranges spent less time hunting than males with prey-poor ranges. In a separate paper (Sodhi 1993b) Sodhi also found that the hunting ranges of both male and female Merlins were smaller if the ranges were prey-rich, and that the majority of males altered the size of their hunting ranges from the incubation to the nestling phase inversely with prey abundance.

While no specific experiments/observations have been carried out on Merlins during the breeding season, the data of Sodhi (2005) is consistent with that found for Eurasian Kestrels in the studies of the Dutch Groningen University group which have already been mentioned. It is therefore considered instructive to set down the Figures (overleaf) which illustrate the findings of the Dutch group as the time budget and weight changes associated with those time budgets are considered to be broadly representative of the behavioural pattern and weight changes of adult breeding Merlins.

Eurasian Merlin nest.
Vladimir Ivanovski.

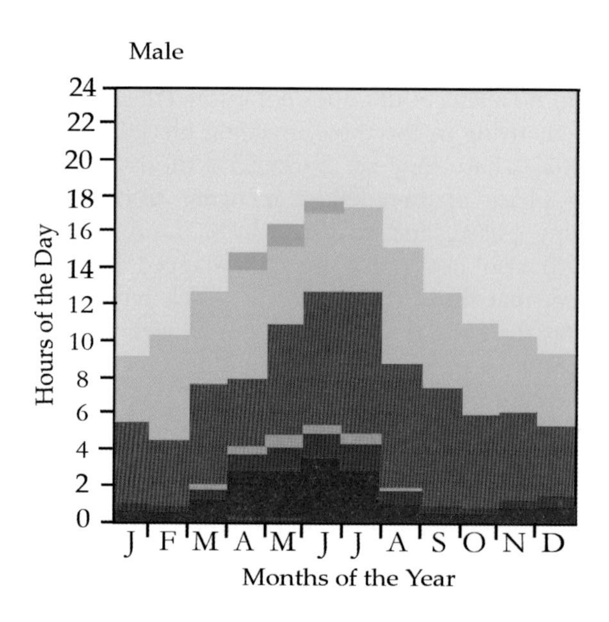
Male
Hours of the Day
24
22
20
18
16
14
12
10
8
6
4
2
0
J F M A M J J A S O N D
Months of the Year

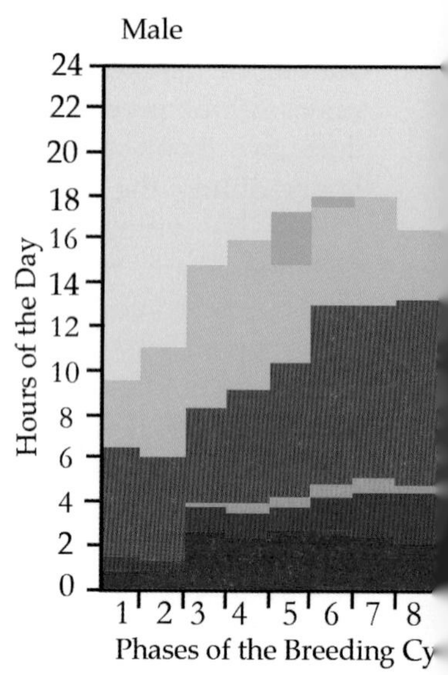
Male
Hours of the Day
24
22
20
18
16
14
12
10
8
6
4
2
0
1 2 3 4 5 6 7 8

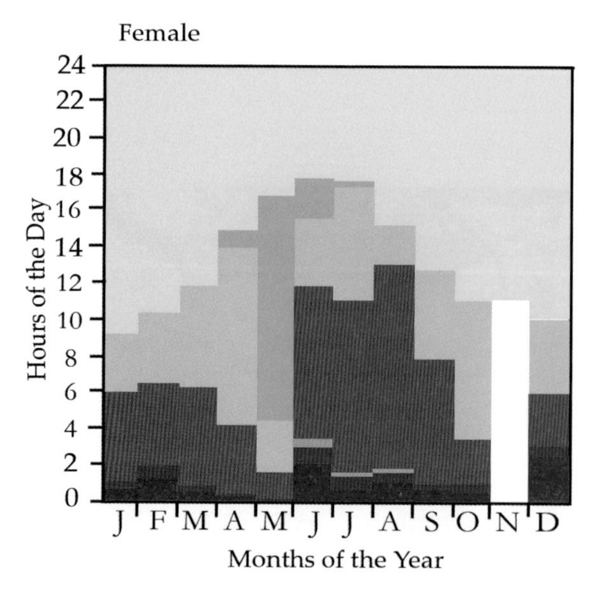
Female
Hours of the Day
24
22
20
18
16
14
12
10
8
6
4
2
0
J F M A M J J A S O N D
Months of the Year

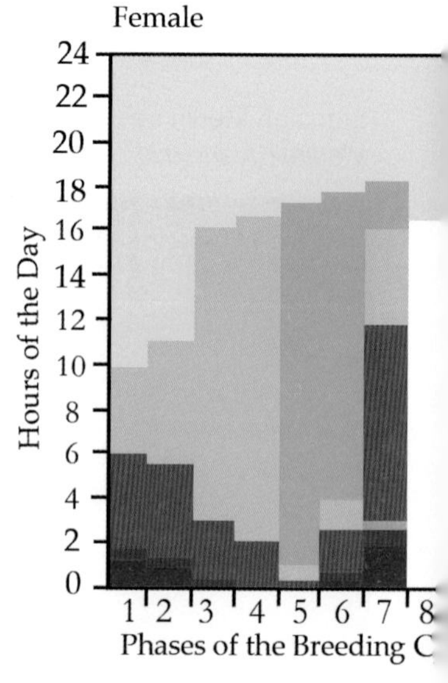
Female
Hours of the Day
24
22
20
18
16
14
12
10
8
6
4
2
0
1 2 3 4 5 6 7

Opposite
Figure 9
Time budget of male and female Kestrels throughout the year (left) and by phase of the breeding cycle (right). Phases of the breeding cycle are:
1. Wintering unpaired
2. Wintering paired
3. Courtship feeding
4. Egg laying
5. Incubation
6. Nestlings below the age of 10 days
7. Nestlings above the age of 10 days
8. Dependent juveniles
9. Post reproductive moult

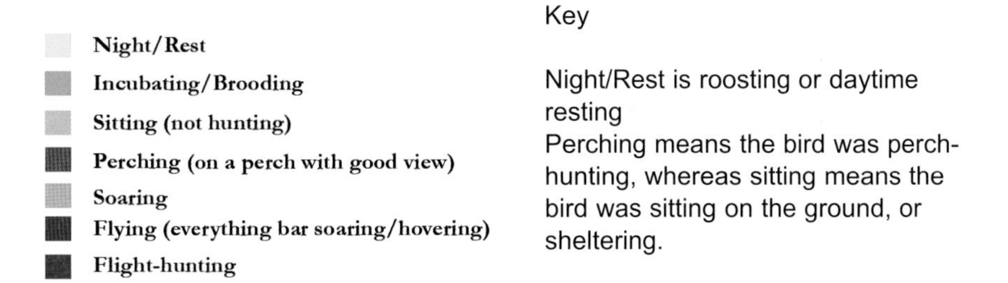

Key

Night/Rest is roosting or daytime resting
Perching means the bird was perch-hunting, whereas sitting means the bird was sitting on the ground, or sheltering.

Redrawn from Masman *et al.* (1988b) © Wiley Online Library.

Below
Figure 10
Variation of daily energy intake and expenditure in male and female Kestrels throughout the year.
Redrawn from Masman *et al.* (1988a).

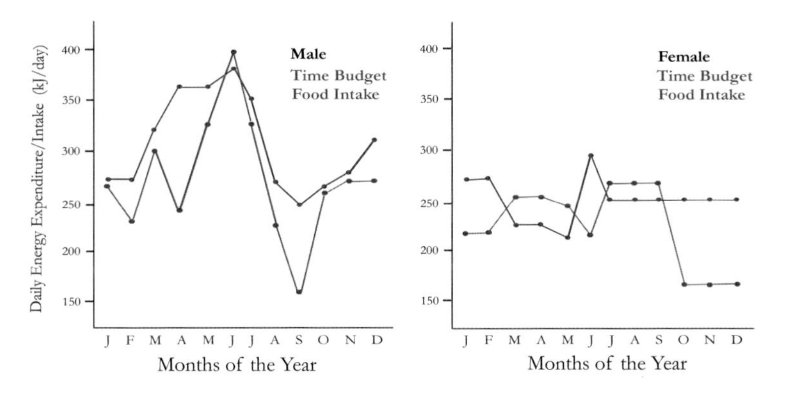

The clear implication of Fig. 10 is that the difference between energy intake and energy expenditure in both male and female Kestrels will result in body weight changes. These are shown in Fig. 11 which indicates significant weight changes during the phases of the breeding cycle.

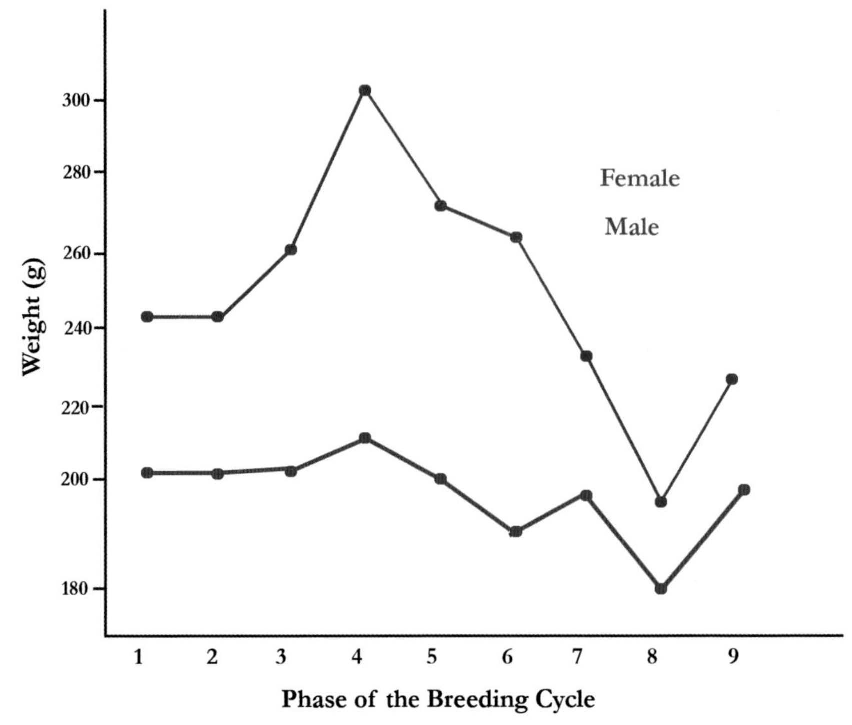

Figure 11
Variation of body weight in male and female Kestrels by phase of the annual life cycle.
Phases of the breeding cycle are as for Fig. 9.
Redrawn from Masman *et al.* (1986).

One interesting outcome of the Dutch group's work on the Eurasian Kestrel was the effect of the male bird's time budget on breeding time. Without the constraints imposed by the physiology of males and females, a pair of animals could breed at any time when the availability of food allowed the female to successfully overcome the energy expenditure of reproduction and the pair could provide sufficient food to raise the resulting offspring. In practice, most animals are 'programmed' to breed at a specific time, the time being defined by the availability of food resources since these are usually not available consistently through the year. Masman (1986) calculated the daily energy expenditure of male

Kestrels during the year (Fig. 12). Masman found that from August to mid-January the shorter northern day meant that the male did not have the time to catch the food required to feed a female and brood as well as feeding himself. In principle Kestrels could breed in the autumn when the vole population is comparable to that of spring, but the male is then constrained by hunting time.

Merlins hunt chiefly in the early morning and late afternoon during both the breeding season and in winter. In their study of urban-dwelling Merlins in Saskatoon, Warkentin and Oliphant (1990) noted that the falcons hunted preferentially at 09.00 and 16.00 (Fig. 13 overleaf) and considered it likely that these times were chosen in order to restore energy levels (in the morning) and to increase energy levels (in the evening) to better survive the long hours of cold when darkness precludes hunting. That reasoning seems entirely plausible, but it is worth noting that in a study on Eurasian Kestrels in Holland (RUG/RIJP, 1982) where the falcons hunt Common Voles, the researchers noted that the voles emerged from their burrows to feed in two hours cycles, and that the arrival of Kestrels to hunt corresponded with these probable emergence times. The number of Kestrels hunting then decreased over the two-hour feeding time as a successful bird would catch prey and fly off to consume

Figure 12
Estimated daily energetic expenditure throughout the year assuming breeding starts at that time. The arrow indicates the average start of courtship of Kestrel pairs.
Redrawn from Masman (1986).

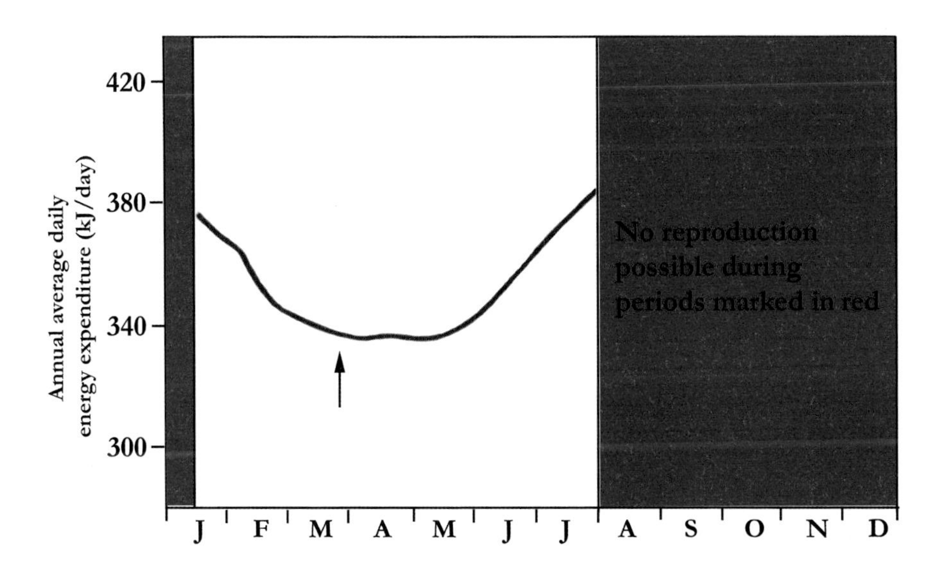

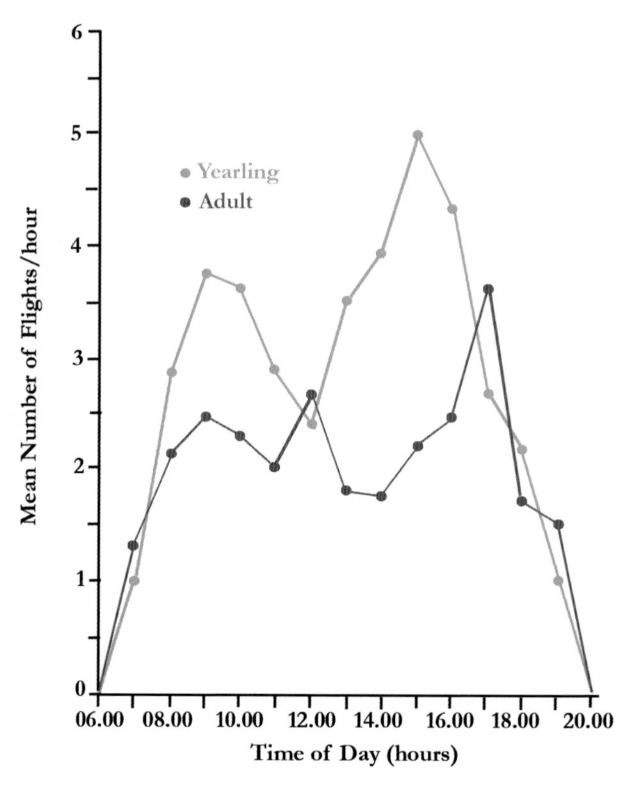

Figure 13
Mean number of flights per hour by adult and first year Merlins in winter. Redrawn from Warkentin *et al.* (1990).

it, perhaps not returning until the next vole feeding period. The Dutch noted that Hen (Northern) Harriers saved about 90 minutes of flight-hunting daily by adopting this temporally linked behaviour (corresponding to about 12% of their daily energy intake). While the main prey of the urban Merlins are primarily foragers with no specific temporal linkage, they are subject to the same environmental stresses as the Merlins and so will likely seek food early in the morning and then prior to sunrise for similar reasons and so are likely to be away from cover. This, and the imperative to feed, will make them more vulnerable at those times. As with other species, the Merlin's prey will also fill their crops, then retire to empty them, and it is likely that the Merlins will adapt their hunting behaviour to coincide so as to reduce their own energy expenditure.

The late afternoon hunting forays noted by Warkentin and Oliphant (1990) are confirmed by Dickson (1993) for Merlins wintering in Galloway, south-west Scotland, (Fig. 14) though the actual peak time, 14.30, differs as a result of the earlier sunset in Scotland: in each case the

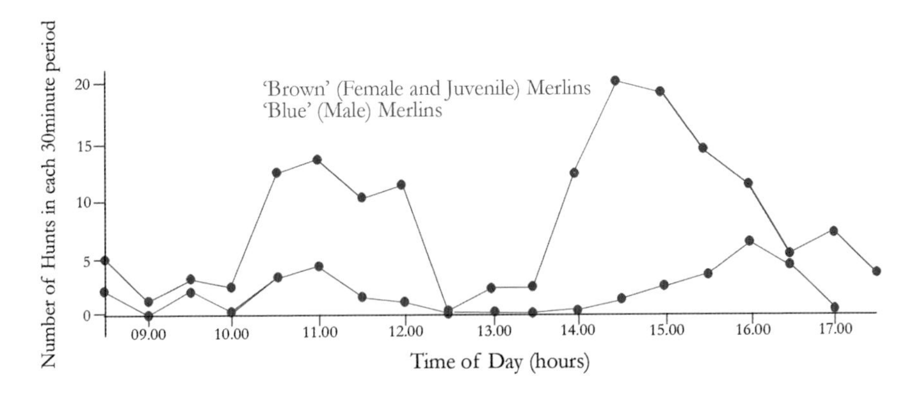

Figure 14
Hunting times of Merlins wintering in Galloway.
Redrawn from Dickson (1993),

hunting peak is about an hour before sunset. However, the morning peak noted by Dickson, 11.00, occurs about 2½ hours after sunrise (and after the Merlins had left their roosts), which is much later than the hunting peak in Saskatoon, which more or less coincides with sunrise. The earlier hunting by the Canadian Merlins is likely to be due to the winter temperatures in Canada being lower, indeed very much lower, than those in Scotland.

These four Eurasian Merlin chicks have been brought down from a tree nest in Scotland for ringing.

Breeding Part 1

In this Chapter we consider the early phase of the breeding cycle, before covering egg laying and chick growth in the following Chapter.

Territory

As with many raptors there is a difference between a pair's territory and their hunting range, the latter often being larger (sometimes much larger), though never smaller. In their study of Asian Merlins, Morozov *et al.* (2013) explicitly point out the difference between the two, noting that the falcons had an area around their nests which was rigorously defended, i.e. within which an intruder would always be attacked. Normally this area would be 200-300m in all directions, but could be as little as 50m in some cases. However, defining territory solely on the basis of intruder distance probably underestimates the actual size, as Merlins will pursue intruders to boundary edges they perceive around their nests and will also attack intruders at greater distances if they come within these boundaries, so that while there may well be a reduced area around the nests in which defensive action is heightened, there will also be a wider area over which defence is also maintained. As we shall see below, defining territories by defensive tactics can be more complicated than it at first appears.

In their study of Merlins in northern England Newton *et al.* (1978) noted that the falcons frequently flew further to hunt than the minimum distance between nests, while in another in two Swedish National Parks (Padjelanta and Stora Sjöfallet), Wiklund and Larsson (1994) found that hunting ranges could be twice the pair's territory, averaging about 2km^2 in some study years. This figure is significantly larger than that quoted by Morozov *et al.* (2013) but this could arise as a result of the noted differing definitions of territory. The Swedish data also suggests a very good hunting area, and in general both hunting ranges and territories are likely to be larger. (The Swedish range is not unique, in a study in the far east of Russia, Shubin (1984) found one hunting range of 2.3km^2 and another of 4km^2.) In studies in North America, the range sizes of male *F.c. richardsonii* were much larger, averaging 21.3km^2 (range 12.6km^2 to 28.1km^2 – but only three territories were identified) in Montana (Becker and Sieg 1987a) and 8.2±2.7km^2 in Saskatoon (Sodhi and Oliphant 1992) as an average for radio-tagged, urban-nesting Merlins. The large standard deviation on the latter range figure arises because of the difference between resident males and those of immigrant males (defined, in this context, as birds which had been hatched in the rural surroundings of Saskatoon, but which had immigrated into the city), the range sizes being 6.3±1.3km^2 and 33.7±12.1km^2 respectively. The very large difference in the male ranges, which is statistically significant (at the 99.5% level) resulted from immigrant birds spending time hunting in rural areas. Of 12 resident males, eight hunted solely within the city and so had smaller ranges, while all the immigrant males spent time in rural areas adjacent to the city, which meant longer flying times and, hence, larger hunting ranges. Two males whose nests were only 0.8km apart reinforced the view that immigrant males often hunted in habitats which corresponded to their natal areas: the resident male never left the city to hunt, while the immigrant spent one-third of its time hunting in adjacent rural areas. For females the hunting ranges were 6.6±3.4km^2 and 8.6±1.6km^2 respectively for resident and immigrant females. While the difference in this case is not statistically significant, the study showed that immigrant females spent more time in adjacent areas than resident females. Clearly the impulse to travel to natal-like areas is less pronounced in female Merlins, presumably because once they have their own brood the imperative is to spend as little time away from the nest as possible.

In a study in south-west Scotland, Orchel (1992) calculated the minimum area of grass and heather moorland required to support a breeding pair of Merlins as 20km^2, in good agreement with the Becker and Sieg (1987a) value for rural Montana. While both figures are higher than that found for the urban-dwelling Merlins of Saskatoon, that would be expected as passerine densities in cities are in general higher than those of rural areas because of deliberate or accidental feeding by human city dwellers.

As a consequence of this differential size, male hunting ranges often overlapped, Sodhi and Oliphant (1992) noted that the overlap percentage was about 11% at all stages of the breeding cycle (i.e. during the incubation, nestling and fledgling phases), but that this figure was the mean of a range of 0% to 77% (based on conservative estimates as not all neighbours in a year were radio-tagged). In general, raptor hunting ranges and territories are much more closely linked during the breeding season and are, for the obvious reason of exclusive maintenance of resources, protected more stridently. But this was not what Sodhi and Oliphant found among the urban-breeding Merlins of Saskatoon. They suggest three principal reasons for this lack of significant conflict as a consequence of territory defence. Firstly, that the high breeding density of Merlins may have made range defence costly in terms of time and energy expenditure; secondly, that local prey was abundant and its population was stable, reducing the potential benefits of exclusive hunting ranges; and thirdly, that the prey (almost exclusively small birds) is mobile and mostly taken in surprise attacks which means that defending a particular area is not worthwhile. Interestingly, a study of *F.c. richardsonii* breeding in south-eastern Montana (Becker and Sieg 1987a) indicated that the falcons did not aggressively defend their hunting ranges during the breeding season. In their study, Becker and Sieg noted that male Merlins were hunting up to 9km from their nest site and that in one case the nest site of one pair of Merlins was located within the home range of another male. Warkentin and Oliphant (1990) in a study of *F.c. richardsonii* wintering in Saskatoon, also noted instances of male Merlins ignoring second birds encroaching on their home ranges, but also saw instances of aggression, which sometimes involved extended periods of chasing of the offending intruder. The relatively sanguine attitude of North American male Merlins is a surprising finding, particularly during the breeding season when one of the primary functions of a male bird is to ensure that that no other male has access to his female partner, and does not seem to be shared by other sub-species.

Wintering British Merlins are also more aggressive than the North American urban Merlins, although in this case there is a logical reason. In his observations of wintering Merlins in eastern Scotland, Cresswell (1996) noted that in 30 observations of two Merlins on 97% of occasions one bird was chasing another. Cresswell suggests that as Merlins are essentially stealth hunters, using surprise as a technique, the second Merlin would be an intrusion, potentially alerting prey to the presence of a raptor and so reducing the likelihood of a kill. In urban Canada where the prey density is higher such a protective imperative does not apply to the same degree.

Pair Formation

Merlins are considered to be monogamous, or largely so, though both Newton (1979) and Cosnette (1991) in Britain and Sodhi (1989) in North America present data suggesting that polygyny occurs. Newton reports the observation of another who saw a male Merlin visiting more than one nest site with no other male apparently present. Cosnette (1991), in Scotland, saw two females with nests close together with only one male Merlin apparently in the area. Later a 'combined' group of three juveniles was seen together, one about eight days younger than the other two, which were being fed by both females with no signs of aggression between them: the single male was seen passing prey to one of the females. In a study of urban-dwelling *F.c. richardsonii* in Saskatoon Sodhi (1989) watched a male Merlin (colour-marked so there was no doubting that it was the same bird) maintain two nest sites 450m apart, making prey transfers to each and copulating with the females at the different sites. However, the attempt to raise two families failed: the male apparently deserted one nest, and the female deserted a few days later. At the other nest four chicks were successfully raised. The result is interesting in the light of the hypothesis of Korpimäki (1988) regarding the potential polygyny of European raptors. Korpimäki suggested that polygyny was more likely in situations where prey was abundant, pair bonding was annual rather than long-term, and territoriality was weak. On the basis of this hypothesis Korpimäki suggested that polygyny was more likely in rodent-eating birds as rodent populations tend to be more cyclic, with well-defined peaks; and that polygyny would therefore increase in northern-breeding raptors as rodent population peaks tend to be more pronounced in northern climes. In his study of Eurasian Kestrels, a northern, rodent-eating species, that is exactly what Korpimäki found. As prey density in urban areas is in general higher than in rural areas, the finding of Sodhi is therefore in accord with Korpimäki's hypothesis, though the fact that Merlins are bird-feeders would act against polygyny in the more usual habitat of the species. Since Wiklund (2001) showed a correlation between rodent density and the Merlin breeding population, it would be interesting to see if Merlin polygyny increased during peak rodent years in northern Fennoscandia.

There is also evidence of extra-pair copulation (see *Copulation* below).

Both first year male and female Merlins can breed, but in a study in north-east Scotland using DNA profiling based on the extraction of DNA from feathers found at nest sites to identify individual falcons, Marsden (2002) found that while females often did breed in their first year, males did not. Marsden found the average age of first breeding was 1.3±0.1 years for females and 2.3±0.1 years for males. These figures are consistent with those from a study of urban-dwelling Richardson's Merlins in Saskatoon, in which Lieske *et al.* (1997) calculated the average age of Merlins at first breeding was 1.9±0.7 years for males, and 1.3±0.6 years for females (as percentages of the population, Lieske and co-workers found that 27%-37% of males bred in their first year, but almost all had bred by their second year: for females, about 80% bred in their first year). Lieske and co-workers noted no difference in nest failures as a consequence of the age of either of the breeding pair. The reasons why females breed earlier than males is not well understood. It has been proposed that this might reflect higher mortality in females (the differential mortality hypothesis, though this could apply to both males and females), or that male birds require greater experience before breeding, reflecting the greater burden placed on them in terms of the need to hunt in order to provision a female during the egg laying and incubation phase of breeding, and in the provisioning of growing nestlings. This second hypothesis could also apply to both males and females, but given the very large difference in the energy investment of males in the production of offspring (a trait which male Merlins share with other male raptors) it is more likely to be seen in males. The study of Lieske *et al.* which provided the data on the age of first breeding was carried out on urban-dwelling Merlins in Saskatoon, Canada, and there is evidence from the same city (see the observations of James and Oliphant (1986) in *Copulation* below) that first-year males may sometimes aid breeding adult pairs, supporting the idea that in some cases experience is required before successful breeding can be attempted.

Building on the study of Lieske *et al.* (1997) with the same urban-breeding population, Espie *et al.* (2000) attempted to explore the reasons for differential breeding in male and female Merlins by considering the brood size of birds of differing ages, the hatch dates of the chicks, and the total number of fledglings produced by birds over their lifetime (the Lifetime Reproductive Success – LRS). Espie and co-workers also studied the number of young produced by individual birds which survived and returned to the study area. The results indicated that brood size was positively correlated with the age of parent birds, though the increase was more pronounced for males than for females. However, what was interesting was that the increase was apparent only for birds younger than the mean age of the population: for birds older than the mean age the brood size declined. In other words, birds which survived to breed again had larger broods until they reached the mean age, but after that they had smaller broods as these increased the likelihood of survival to breed again. The results also indicated that, for older parents, whether they be male or female, the hatch date of their chicks was earlier than that of their younger counterparts; and that, as would be expected, LRS was positively correlated with age for both sexes. As regards the number of offspring which returned to the study area, the result was no better than chance for male Merlins, but female Merlins produced no offspring which returned.

Considering their results, Espie *et al.* (2000) concluded that there was support for the differential mortality hypothesis for both sexes, though support was much more pronounced for females. There was also support for the experience hypothesis, particularly for male Merlins, a finding which is supported by the observation of James and Oliphant (1986) – see *Copulation* below. These results are, perhaps, to be expected. Males have to provision the female and the brood, which means hard work and requires hunting skills: the former might take its toll on the bird, the latter needs experience which might mean a year spent acquiring skills before mating. Females must acquire the bodily reserves for egg laying and, later in the breeding cycle, must aid provisioning a growing brood: both will take their toll on the bird, though neither will prevent a female from mating in her first year as lack of hunting skills will be alleviated by a good partner during the latter stages of brood raising. Overall, as would definitely be

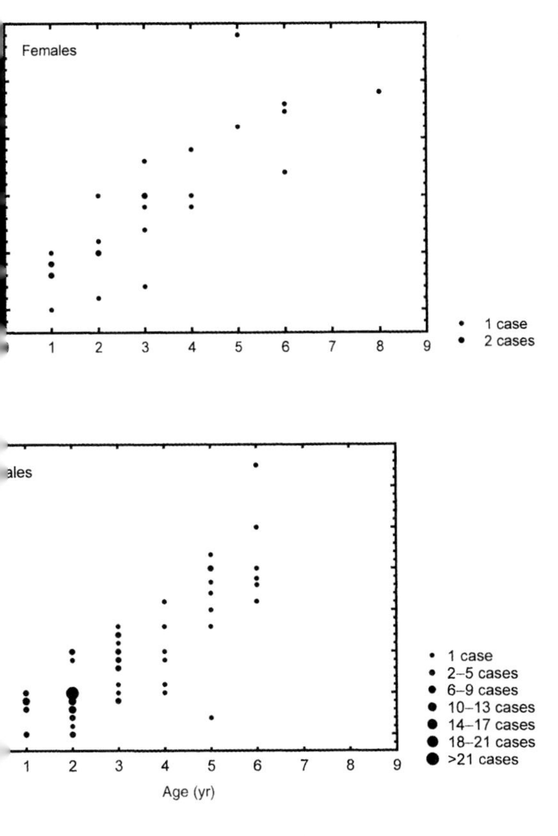

Figure 15
Scatter diagram of age attained and reproductive success of male and female Merlins.
Reproduced from Espie *et al.* (2000).

expected, the longer a Merlin lives the more chicks it will successfully raise (Fig. 15), i.e. the higher the LRS. Espie *et al.* (2000) found that a female will produce a mean of 9.2±6.2 fledglings in her lifetime, a male will produce 7.4±5.9. The reduction in male LRS is a result of losing a year to gain experience: male Merlins are trading loss of first year breeding for long term reproductive success. The LRS for the female Merlins was higher than that calculated in a study in northern Sweden where Wiklund (1995) measured 6.4±4.6, with a range of 2 to 24. While the high standard deviations in both cases mean that there is considerable overlap in the two LRS distributions, the higher value for the urban-dwelling Merlins probably relates to the reduced nest predation of that population.

The differential age at first breeding between the sexes noted by Lieske *et al.* (1997) is consistent with the studies of Wiklund (1996) and with the studies of other small falcons, which have shown that first-year females were more likely to be paired with adult males than were first-year males with adult females. In another study of pair formation (Warkentin *et al.* 1992a) it was found that adult-adult and first year-first year pairs were formed more frequently than would be expected by chance, supporting the contention that adult birds seek to pair with other adults as the breeding experience each brings to the partnership enhances breeding success. The Warkentin *et al.* (1992a) study (on urban-breeding Merlins) did not support the idea that first year females were more likely to breed with adult males than first year males were with adult females, finding as many first year male-adult female pairs as the reverse: 10% of breeding pairs included a first year male, 12% of first year included a female. However, as the data on such couplings were limited this result does not provide evidence to contest the Wiklund suggestion that such adult-young bird pairs favour females. Newton *et al.* (1986) also studied the number of first year Merlins breeding in Northumberland, north-east England. Their data supported the idea that first year females were more likely to breed, 18% of pairs including a first year female in a study which lasted across nine years, while only 8-9% of pairs included a first year male over the same period.

Several reasons have been suggested for this assortative mating. One suggests that adult males arrive earlier at the breeding grounds than first year males and so are able to establish territories that are richer in food resources, and it is certainly the case that adult-adult pairs tend to breed earlier than pairs in which one bird is a first year, and that earlier breeding usually results in a higher number of fledglings. By contrast, later breeders, which often means pairs which are either both first years or include one first year, have lower reproductive success: Warkentin *et al.* (1992a) noted that adult-adult pairs fledged on average 4.2±1.2 young, while pairs with at least one first year bird, bred later and produced 2.6±1.6 fledglings.

Female Merlins may recognise the better potential of a territory in terms of provisioning herself and her brood and choose appropriately. Females may also recognise better hunting abilities in the males themselves, the study of Warkentin *et al.* (1992a) suggesting that females might choose males on the basis of longer tail length, longer tails having been shown to influence the hunting ability of raptors chasing avian prey. Since first year males tend to be smaller than those of the adult males, and with shorter tails, the females' preference for a longer tail might also count against the younger males. Warkentin and co-workers also found that females tended to choose males with the same wing chord length in successive years. In his work Espie (2000) found that wing chord length was the best single measure of body size in Merlins. Espie confirmed the result of Warkentin *et al.* (1992a) that females tend to choose the same size mates in successive years, despite finding that body size actually had minimal influence on breeding performance for either sex: Espie found no repeatability in terms of male choice of mate.

Although more work is required before final conclusions can be drawn, given the significantly larger number of adult-adult pairs, it is perhaps a surprise that first year birds manage to find partners at all, but any population of birds will inevitably include a number which do not readily find a mate, and for these it is probably better to choose a bird of reduced 'quality' (without defining exactly what 'quality' means in this context, but noting it might be territorial holding, apparent hunting fitness, experience, or a combination of these and other factors) than not to attempt to breed at all. For a male bird, given that much of the burden of provisioning mate and brood falls to him, there is less risk in choosing a first year female than there is for a female who will be much more dependent on her chosen male. Consistent with this idea the study of Warkentin *et al.* (1992a), showed that the lowest productivity (as measured by number of young fledged) was seen in first year male-adult female pairs, a finding which was also supported by the study of Newton *et al.* (1986) which saw more than double the number of first year females breeding than first year males.

With the proviso that adult birds prefer to mate with adult birds, and that female Merlins may choose males on the basis of body size as a means of providing evidence of hunting ability, mate selection appears random, with little evidence to suggest pair fidelity (see, also, *Chapter 8: Breeding Territory and Mate Fidelity*). James *et al.* (1987b) observed a mating between a female Merlin (aged four years) and her son (aged two years): the pair's history was known, as they had both been ringed as chicks and, along with 10 other pairs, were subsequently trapped as part of the study. Such a mating is considered evolutionarily disadvantageous, though in this case the pair produced five offspring which was more than the average for the area under study, and if not explicitly avoided, it is rare as a consequence of dispersal after breeding in those species which do not show high mate fidelity. However, the in-breeding fraction defined by the single pair of the 11 trapped (i.e. 9%) was shown to be an overestimate due to the small sample size. A later study of the same urban-breeding Saskatoon Merlin population by Warkentin *et al.* (2013) found only one father-daughter pairing in 35 possible pairings (1 of 35 = 2.8%), two mother-son pairings (2 of 88 = 2.3%), no full sibling pairs (0 of 21) and one half-sibling pairing (1 of 21 = 4.8%): all of the pairings produced numbers of fledglings which were within the expected range and all of which appeared viable.

Warkentin *et al.* (2013) note that it has been suggested that post-breeding dispersal, particularly sex-biased dispersal, is a mechanism that has evolved to avoid in-breeding, but note that in the cases identified the dispersal distances of the individual birds were greater than the mean of the population. It appears, therefore, that the identified pairings were random events. This result, therefore, seems in accord with the study of American Kestrels by Duncan and Bird (1989) who found that females did not discriminate against male siblings in their choice of mate, the authors suggesting that high mortality in, and the limited number of nest sites available to, the species may mean that incest is rare in the wild and so may not influence mate selection, an argument which is equally valid for the Merlin. Since Duncan and Bird were studying captive birds it might also be argued that the females had little choice in mates and so might have exhibited behaviour which was rarer in the wild, though it is known that incestuous relationships exist in other falcons (**Box 16**).

Box 16

Greenwood (1980) noted that while many species of animals are faithful to their natal area, in birds the tendency is for males to be more philopatric than females while the reverse is true for many mammals. But either way, the fact that the degree of philopatry differs in the two sexes reduces the likelihood of incest, which is clearly beneficial to the species. There is also evidence that female birds prefer philopatric males because occupation of a territory suggests both a willingness to defend it and an ability to withstand competition, and so points to the likelihood that resources will be available for breeding. Knowledge of a territory also implies some understanding of local predators which is also advantageous. These issues were neatly illustrated in an experiment in central Finland (Hakkarainen and Korpimäki 1996) with three owl species – Eagle Owl, Ural Owl (*Strix uralensis*) and Tengmalm's Owl (*Aegolius funereus*). Of the three, Tengmalm's Owl is by far the smallest and is predated by the larger two. However, the larger two owls cannot access the small tree holes used by breeding Tengmalm's Owls to take their nestlings. Tengmalm's Owls are also out-competed by Ural Owls for the small rodents on which both prey. Eagle Owls take much larger prey and so are not direct competitors for resources. By erecting nest boxes in Eagle and Ural owl territories Hakkarainen and Korpimäki were able to show that the breeding of Tengmalm's Owls was less affected in Eagle Owl territories than in those of Ural Owls where the smaller owls suffered both predation and competition for prey. Most breeding attempts of Tengmalm's Owls near Ural Owl territories failed during courtship and those that succeeded saw clutches laid 11 days later which would have resulted in lower fledgling rates. Tengmalm's Owl pairs close to Ural Owls invariably consisted of younger males and females: experienced males were taking the best sites and experienced females were recognising the advantages they offered.

Before leaving mate selection, it is worth recalling that juvenile male Merlins are very similar to females in plumage, a trait which is mirrored in, for instance, the Eurasian Kestrel and which has led to a discussion of whether the coloration endows the young male with a reproductive advantage. Hakkarainen *et al.* (1993) noted that a similar phenomenon was found in many dichromatic species during their first potential breeding year, and that two hypotheses have been put forward as an explanation. In the female-mimicry hypothesis it is conjectured that the mimicry aids the juvenile males to mate by deceiving adult males into believing that they are females and, therefore, not competitors for mates.

In the status-signalling hypothesis the mimicry allows adult males to differentiate low status males, again reducing competition. This second hypothesis assumes that adult males can distinguish the sex of young males. Hakkarainen and co-workers noted, in a study in Finland, that juvenile Eurasian Kestrel males tended to choose breeding sites closer to adult males and that this increased both their breeding success and the outcome success of the breeding, particularly if they mated with adult females. This suggested that plumage mimicry was an adaptive feature to enhance breeding potential, but still did not differentiate between the two hypotheses. To aid with distinguishing the two, the Finnish team showed captive adult male Kestrels both adult female and female-like juvenile males. If the two were shown simultaneously the adult males preferred the females, but if the two were shown separately the adult males were unable to distinguish them. This strongly suggests that adult male Kestrels are not good at distinguishing sex, providing good evidence for the female-mimicry hypothesis.

Displays

More display flights have been identified in North American Merlins than in those of Eurasia. Indeed, the work of Feldsine and Oliphant (1985), based on observations of *F.c. richardsonii* in Saskatoon, Canada, is unsympathetic towards the comment by Cramp and Simmons (1980) that display flights (of Eurasian Merlins) are 'rather inconspicuous and rarely observed', noting that 'the Merlins we observed regularly performed complex aerial displays that were nothing short of awesome'. Working on urban-dwelling Merlins in Saskatoon, Feldsine and Oliphant noted a whole series of display flights, many of which (though not all) have been observed by the present author, not only in North America, but in Iceland, Britain and Fennoscandia.

At the start of the breeding season male display flights are used to define and advertise their territory, both to rival males and potential mates. The main display flight by the male at this time is a fast flight with strong wing beats during which he rocks from side to side in order to flash his underwings, making occasional '*chip*' calls. Feldsine and Oliphant (1985) call this 'power flying'. During this display the male may also dive steeply and quickly, pulling out of the dive in a sharp upward turn. Feldsine and Oliphant call this 'power diving', and also note a rocking glide, which is power flying, but without the strong wing beats.

British observers report that once a female has become interested the male makes conspicuous short flights from perch to perch around his territory, frequently making the '*kee-kee-kee*' call, presumably advertising that he possesses a territory of sufficient size and prey potential to be worth considering. This flight seems a variation on the Feldsine and

Oliphant power flying. The male may also make flights in the direction of possible nest sites, often hovering above the site or making slow circles above it, again presumably to impress potential mates. Morozov *et al.* (2013) note this display in Asian Merlins, suggesting the male makes pirouettes above potential sites. Feldsine and Oliphant (1985) do not record this display explicitly, but do define a slow landing display, which has many of the same components, though the Canadians define this as a pre-copulation display rather than as a flight which defines a potential nest site. Once a female has shown an interest the two birds may chase around the territory, each taking a turn at being chaser or chased (**Box 17**): Feldsine and Oliphant refer to this as 'tail chasing'. The two will then perform the main courtship display, which is similar to that of other small falcons, the two birds circling high above the proposed nest site with a specific wing action. This was delightfully termed *zitterflug* – trembling flight – by Tinbergen (1940) in his work on the Eurasian Kestrel, and has occasionally been called 'shivering flight': Feldsine and Oliphant refer to it as 'flutter flight'. In this display the birds fly slowly while the wings vibrate quickly. The male usually interrupts shivering wing beats with slower beats and makes the '*chip*' call, the female often replying with the standard '*kee-kee-kee*', or she may make a curious whistling call.
Feldsine and Oliphant (1985) also note a high soaring display in which one or several Merlins circled at heights of several hundred metres, then set their wings and soared if conditions allowed. Though the Canadians note having seen soaring often, it has not been reported elsewhere, nor seen by the present author. It is possible that soaring may require conditions which occur more frequently over the Canadian prairies.

The passing of food between male and female is the final stage of pair formation, the female makes the begging call to her mate to initiate food passes (**Box 18**). Food passes may take place in the air, but more frequently are made at a perch, usually foot to foot, but occasionally bill to bill. Interestingly, Simms (1975) reports seeing male and female Merlins

Box 17
In a communal roost in Scotland Dickson (1973) noted that 2 or 3 birds, usually a male with one or two females or juveniles would frequently engage in aerial chases soon after leaving the roost, or in the evening before returning to the roost. Keeping close together the birds would climb high, then dive through the wooded roost thicket, before repeating the performance, all the time calling loudly. Occasionally at the top of a climb two birds would stall and briefly touch talons. The display lasted anything from a minute or two, to almost 30 minutes. Dickson, almost certainly correctly, considered the activity an exercise in pair bonding.

Box 18
Nelson (1970) feels the term 'courtship feeding' is not appropriate to describe the exchange of food between male and female. Although in the early stages of courtship it is likely that the giving and accepting of food cements the pair bond, Nelson considers that continued instances of food passing are related just as much to the fact that the two birds are ensuring that the female acquires the reserves necessary for egg laying and, consequently, reproductive success as they are to bonding. Therefore, it is in the interests of both that the male passes food to his mate after the 'courtship' phase.

touching talons during what appeared to be food passes, but in which no food was involved. This behaviour, which Simms observed several times over a period of years, suggests that the birds may use 'food passing' close approaches to reinforce the pair bond even if no food is actually involved. Following the food passing phase, nest selection takes place.

Although the male clearly prospects suitable nest sites and invites the female to view them, it is not clear whether to not he actually chooses a site or just makes suggestions. Nest viewing is usually accompanied by a great deal of calling by both birds. Often the female will take the brooding position at several sites (Nethersole-Thompson and Nethersole-Thompson 1943) before a final decision is made, that decision seemingly the female's alone. Once the decision is made the pair may then reinforce their bond by making displays at the nest, the male making the '*chip*' call while standing at the nest with a fanned tail and drooping wings. The birds may also bow to each other at the nest site or engage in bill nibbling. Feldsine and Oliphant (1985) also note the male's nest display: in a captive pair of Merlins the male took up an incubation pose on the nest before extending drooping wings and raising his tail vertically while the female watched passively.

Nest Sites

As with all falcons, Merlins do not construct nests, utilising the stick nests of other species or a ledge on a cliff. On the latter the female will excavate a shallow scrape in the substrate. In areas where the birds are ground nesters, a shallow scrape will also be excavated and in Britain, where such nests are located among dense heather or other vegetation, the female will occasionally collect small pieces to form a sparse lining. While such nesting seems, at first glance, to have been due to the deforestation of Britain, following the first agrarian revolution in the Palaeolithic/Neolithic period, and the second revolution around the time of the Industrial Revolution, it is worth noting that the northern British

archipelagos of Orkney and Shetland were never forested to any extent, so the Merlins' catholic nesting habits were not wholly forced by changes in land usage but, rather, imposed by the nature of their northern habitat.

In North America almost all Merlin nest sites are in old stick nests, though in a study in rural Saskatchewan, Fox (1964) noted that while 52% of Merlins were found using the nests of American Crow and 42% to be utilising those of Black-billed Magpie, 8% were in tree cavities. Trimble (1975) reports Merlins nesting under the roofs of deserted buildings and on cliff ledges. Studying Merlin nest site selection Warkentin and James (1988) noted that over much of its Nearctic range the species nested in old corvid or raptor stick nests which were, predominantly, in conifer trees, though they noted that *F.c. richardsonii*, more frequently nested in deciduous trees. This change probably reflects tree species availability for the original nest constructors: urban–breeding Merlins utilise corvid stick nests constructed in any available tree in recreational areas, cemeteries and schools, but also in residential areas.

In one such city, Saskatoon, where the study of Warkentin and James was made the two researchers conducted surveys of the city to identify Merlin nests and then considered the selection of them against various criteria – tree species, height and diameter of tree, height of nest, distance to nearest building and so on – in order to assess what the Merlin pair were seeking for their ideal brood site.

Warkentin and James found a total of 146 utilised corvid nests, 93 of which had been constructed by American Crows, the remainder by Black-billed Magpies. 130 (89%) of the nests were in conifers. The Merlins selected only those nests which were in conifers, and of 58 Merlin nests 54 (93.1%) were in crow, rather than magpie, nests. Crow nests tend to be constructed higher within the chosen tree as well as being significantly higher (8.9±2.9m) above the ground than those of the magpies (6.5±2.0m). Considering the standard profile of coniferous and deciduous trees, it is clear that Merlins prefer trees which offer greater vegetation coverage, and nests which offer greater protection against predation.

The preferential use of crow nests in Saskatoon is interesting as it differs from the preferred use of magpie nests in rural-dwelling Merlins where the falcons utilise nests built in shelter belts and trees in river valleys. In a study in south-eastern Montana Sieg and Becker (1990) found that the falcons were preferentially using magpie nests. However, their analysis points to a logical reason. In the study area the Merlins were

Opposite
Tree nest of Eurasian Merlins in an old corvid nest, Scotland.
Graham Rebecca.

Tree nest of Eurasian Merlins in a basket originally set up for owls, Scotland.

primarily using magpie nests in Ponderosa Pine. The Black-billed Magpie constructs a domed nest, and while the pines are coniferous, they have a very open structure. It therefore seems likely that the Merlins were choosing magpie nests because the dome offered protection from avian predators, the tree itself providing minimal cover. Sieg and Becker also analysed choices in terms of nest height and outlook, and tree position. The falcons were choosing nests with a south-facing aspect which were higher in a tree rooted in flatter ground, and set away from domestic buildings. The chosen tree was usually in a group, but a small group, open enough so that there was easy access and a relatively unrestricted view. The group would be set close to open land over which the Merlins hunted, and the surrounding trees in the group were usually higher, which would offer further protection against avian predators. Since the study of Sieg and Becker (1990) noted Merlin preference for magpie nests in coniferous Ponderosa Pine, the same preference would be expected to exist when the falcons were nesting in country in which deciduous trees were predominant, i.e. the shelter belts of open prairie, since deciduous trees also offer reduced vegetation cover. However, Fox (1964) noted that in rural Saskatchewan where nests in deciduous trees accounted for 84% of the total tree nesting, American Crow nests represented 57% of stick nests, the remainder being nests of Black-billed Magpies. The majority of the utilised stick nests were at heights of 3.5-7.3m above the ground, the

Merlins usually choosing trees which were bare of branches for the first 2.5-3.5m of trunk. However, a preference for magpie nests was also shown by *F.c. columbarius* nesting in White Spruce in the Denali National Park, Alaska (Laing 1985) – though it is worth noting that this study included only a small number of nests – it therefore seems that Merlin nest selection is rather more complex than it might at first appear.

Black Merlins (*F.c. suckleyi*) nest in trees in coastal areas and close to rivers, but also in scattered forest areas of adjacent uplands. From personal observation, but limited sample size, those Merlins (*F.c. columbarius*), few in number, which nest above the treeline in North America are, inevitably, ground nesting on the tundra, despite the vulnerability to predation by Arctic Fox and mustelids.

In Iceland there are fewer trees (in many areas none at all) and only one resident corvid, the Common Raven (*Corvus corax*). Merlins will utilise Raven stick nests where these are available on the ledges, or in the crevices, of cliffs, but otherwise nests on, usually vegetated, ledges on cliffs, lava extrusions, or on steep slopes, at elevations of up to 500m above sea level. Slope nests would be easily accessible for terrestrial predators, but Iceland has only one – the Arctic Fox, *Alopex lagopus* – the population of which was significantly reduced when a bounty was paid for their destruction during the period when the gathering of eiderdown from the nest of the Common Eider (*Somateria mollissima*) was a major industry. Nielsen (1986) notes that the choice of nest site of Icelandic Merlins is consistent with the hypothesis that selection of both territory and nest site by Merlins has been affected by the fact that the falcon shares the landscape with Ravens and Gyrfalcons, the former being a significant predator of Merlin nests. Mutual intolerance of the larger, dominant species has meant that Merlins do not choose nest sites which are favoured by Ravens and Gyrfalcons. The latter nest earlier than Merlins and in general choose cliff ledges which are sheltered by overhangs, which offer protection against predation, but, perhaps more importantly, are snow-free and offer protection against the weather. The later-nesting Merlins are less likely to encounter snow and poor weather and so can accept more open nest sites at the top of cliffs or on the steep slopes below them. However, despite Merlins avoiding Gyrfalcon areas, one volcanic cleft was found (pers. obs.) in which the two species were nesting within 200-300m.

In Britain and Ireland some Merlins do breed in old stick nests, on isolated boulders and on cliff ledges, but despite the existence of terrestrial predators (Red Fox (*Vulpes vulpes*), Eurasian Badger (*Meles meles*) and various mustelids) most nests are on the ground, usually a scrape made in the soil amongst thick vegetation on sloping ground, particularly towards the heads of small valleys, which allow the incubating bird an open view. One fascinating variation occurs in the Orkney Islands, off Scotland's northern coast. The Orkneys are free of mammalian predators (apart from feral cats, domestic dogs and the occasional Eurasian Otter (*Lutra lutra*)), and Hooded Crows (*Corvus cornix*) occasionally build stick nests on the ground which Merlins have been known to use, a combination which now appears unique to the islands. Formerly, in Wales, Merlins also nested on coastal sand dunes at the base of Marram Grass (*Ammophila arenaria*) tussocks: such nests were relatively common at one time but are infrequent/no longer found now.

Across most of Britain nests are on upland moor. In a study in Northumberland, Newton *et al.* (1978) found that of 91 observed nests 25.3% were in trees, 6.6% on crag or boulder tops, 3.3% on cliff ledges and 64.8% on the ground, where nests were concealed among the abundant Heather (*Calluna vulgaris*): although British Merlins avoid nesting among Bracken Fern (*Pteridium aquilinum*). Of the ground sites, 83% were on sloping ground. Newton and co-workers noted that 94% of tree nests fledged at least one nestling as did 75% of boulder top nests. What is, at first sight, surprising, is the fledging success rate they recorded of ground nesting: 68% fledged at least one nestling on sloping ground, and 78% resulted in the same outcome on flat ground. In Wales Williams and Parr (1995) found that the breeding success of ground nests was higher than that for tree nests. In 1993 ground nests had a mean clutch of 4.3±0.5 eggs, which produced a mean brood size of 3.3±0.7 nestlings and 2.0±1.0

Ground nest of Eurasian Merlins, northern England.

fledglings. The figures for tree nests were 3.9±0.7 eggs producing 3.2±1.0 nestlings and 1.5±0.5 fledglings. Though the reason for either complete or partial nest failure was not known in the majority of cases, where it was known (30%) predation was the cause. Four ground nests were predated by Foxes, one by an avian predator, but while no tree nests succumbed to a Fox, six fell to avian predators, principally Goshawk and Tawny Owl.

The ground nest success data are curious as it would be assumed that ground sites are more vulnerable to all predators, but as many Merlin nest on moorland managed for grouse shooting, terrestrial predators are controlled and so have lower population densities and, consequently, pose a lower risk. In principle control of raptors by shooting or poisoning is illegal, but sadly, does continue and Merlins are likely to suffer as a consequence: in reality the risk Merlins pose to adult game birds is minimal, and while they would doubtless take chicks if the opportunity arose, the numbers involved are probably small.

Ground nest do, though, suffer a predation risk and may also be trampled by sheep. But despite these risks Merlins continue to prefer ground nesting. This may be due in part to the protection offered by thick vegetation which ensures that the nests are well-hidden and by the remoteness of sites, factors which may go some way to offsetting this vulnerability. Rebecca (2011) in a study of 717 nest sites from across Britain in 1993/1994 found 78% were on the ground, with 20% in stick nests in trees and 2% on rock outcrops. Of the stick nests, most were in tall conifers, with only 1% in deciduous trees. More recently Merlins have been using artificial nests placed in trees. These numbers would suggest that Merlins might be preferentially choosing ground nest sites in Britain rather than having ground nesting imposed on them by the lack of available trees in preferred hunting areas. However, in his work in south-west Scotland Orchel (1992) noted that conifer planting had altered the nesting habits of the Merlins there. In the period 1960-1976 30 of 31 Merlin nests were on the ground (97%), but as planting accelerated the percentage changed. In 1977-1985 only 71% were on the ground, the remainder in trees, but by 1986-1987 only 15% were on the ground, with 85% in trees.

Wright (2005) found that most of his nest sites were at altitudes of 300-350m, but some were above 500m, which in British terms is high, and none were below about 200m. In Wales some Merlins nest to 650m (Williams and Parr 1995), while the falcons nest up to 800m in the Grampian peaks of north-east Scotland (Rebecca *et al.* 1992). In a study of nest sites across Britain in 1993/1994 Rebecca (2011) found that Merlins nested across an altitudinal range of 10-650m, both the highest and lowest nests being found in Scotland. The British data are therefore consistent with that from Iceland in terms of the elevation.

Ireland's Merlins show a similar elevation ceiling. In his study of the falcons in the Wicklow Mountains McElheron (2005) found nest sites varied in elevation from about 220m to just over 500m. McElheron found that Merlins were almost exclusively utilising the nests of Hooded Crows, which preferentially build in oaks or Scots Pine (*Pinus sylvestris*) in mixed woodland, or Sitka Spruce (*Picea sitchensis*) in conifer plantations. Ground nesting sites were known, but were in a minority (<10%).

In Fennoscandia and across Siberia, Merlins breed in areas which combine the aspects of North American and Icelandic/British habitats, i.e. boreal (the Eurasian taiga) provided there are open clearances, for example the forest edge, or tundra close to the treeline, with a particular preference for river valleys and bogs. In bog areas, Merlins may nest on bog hummocks, especially if Meadow Pipit numbers are high. In Fennoscandia Merlins also breed in coastal areas.

In Finnmark, of 250 nests studied in the period 1982-1993 Tømmeraas (1993a) found that 241 (96.4%) were in Hooded Crow nests, 4 in Rough-legged Buzzard, 3 in Magpie and 2 in Raven. In northern Fennoscandia where birds constructing stick nests are scarce, Merlins may have a limited choice of nest sites and are consequently forced to use nests which are not ideally situated, being either close to human habitation (to which Hooded Crows are often drawn) or relatively remote from hunting grounds. Here they may use ground nests, choosing areas of dense scrub. Tømmeraas also notes historical incidents of Merlins utilising the nests of Goshawk (*Accipiter gentilis*), Golden Eagle (*Aquila chrysaetos*) and Osprey, and also use of what had originally been a Raven's nest, but had then been used by a Gyrfalcon. Tømmeraas also notes a 19th century record of Merlins occupying a tree hole excavated by woodpeckers.

In two National Parks in Sweden (Padjelanta and Stora Sjöfallet) Wiklund and Larsson (1994) found that 134 Merlin selected Hooded Crow nests (88.2% of the total nests), with five in Rough-legged Buzzard tree nests, three in Rough-legged Buzzard cliff ledge nests, five on ledges with no prior stick nesting, and seven on the ground. In this Swedish study it was found that not only did Merlins prefer the nests of Hooded Crows, but that they preferred them to be less than two years old (i.e. the Merlins were using it in the same year as it was built (14.9%) (**Box 19**), constructed nests sometimes being abandoned by the crows, or the following year (44.8%), or the year after (19.4%): only 20.1% of nests were more than three years old) and to have not been used previously by other Merlin pairs. The suspected reason for avoiding nests previously used by Merlins was that the original nesting cup would have been lost, the nest base becoming flat as a consequence, and that in such nests there was a risk that eggs or chicks might be knocked out. Wiklund and Larsson also consider that parasite burdens might be higher in nests that had been used by Merlins in

Box 19
It has been tacitly assumed that Merlins utilising newly constructed nests were taking over structures abandoned by their builders. However, Ivanovski (2003) notes that in Belarus a Hooded Crow nest was recorded as having a clutch of four eggs on 17 April 1994, and that a pair of Merlins were displaying nearby. On 22 May the nest held five Merlin eggs and the crow eggs were in fragments at the tree base. The implication of the find is deliberate destruction of the crow clutch in order to take over the nest.

the previous year (see *Chapter 10: Causes of Death*). The Hooded Crow nests utilised by the Merlins were in Mountain Birch (*Betula pubescens czerepanovii*, a sub-species of the White (or Downy) Birch – note that Wiklund and Larsson refer to the species as *Betula pubescens tortuosa*, the scientific name having been changed since their publication). The trees grow to a height of 8.0m on average (range 5.0-12.m) the crow nests being at an average of 5.4m (range 3.0-8.5m): there was a positive correlation between nest height and tree height. From personal observation above the Arctic Circle in northern Sweden, the nests of Rough-legged Buzzards and Hooded Crow as well as Raven which were utilised by Merlins were invariably in Scots Pine set among the extensive birch forests, rather than in the birches, though in sub-Arctic areas it is more often that available stick nests are in birches. In a later study in the Swedish Arctic-fringe National Parks, Wiklund (1995) found one nest at an elevation of 720m, though all the remaining identified nests were below 640m.

In high latitudes where trees are scarce, Merlins will also nest on the ground on heathland usually dominated by Crowberry (*Empetrum nigrum*) and Dwarf Cornel (*Cornus suecica*). Hills (1980) records a ground nest in northern Finland which was concealed under a juniper at the base of birch in thick woodland. Despite the site being on a steep, east-facing slope, the female Merlin's field of view was extremely restricted – the reason Hills thought the site worth mentioning was because of this, as Merlins in Finland (and elsewhere) invariably choose sites that offer extensive local views.

Morozov *et al.* (2013) studying the nesting habits of the Merlin sub-species in eastern Europe and Asia note that the northern falcons breeding on the tundra and shrub-tundra may use old Rough-legged Buzzard nests on rocky outcrops in river valleys, or may ground nest in thickets of *Juniperus sibirica*, Dwarf Birch (*Betula nana*) or willow (Salix spp.). One nest was found on the roof of an abandoned house. Corvid nests on electricity transmission towers (pylons) have also been used. In the Polar Urals, Merlins use old nests on rock outcrops or nest on the ground and in the forest-tundra zone, the falcons use available stick nests or nest on the

In Eurasia Merlins will readily breed in artificial nest. Most are baskets (as in the photo on p152), but decorous structures have been used, such as this old bucket in Belarus.
Vladimir Ivanovski.

ground. In the southern Yamal Peninsula, of 121 nests, 112 (92.6%) were in stick nests in trees, the remaining nine on the ground. Of the stick nests 103 were of Hooded Crows (92% of all stick nests), with six of Rough-legged Buzzard and three of White-tailed Eagle (*Haliaeetus albicilla*). On one occasion one Merlin pair was found to be breeding in an old Fieldfare nest (V. N. Kalyakin in Morozov *et al.*) which, it would be assumed, was barely big enough for the clutch and incubating female, let alone pre-fledge juveniles. Of course, in areas of the forest-tundra where trees and, therefore, stick nests, are scarce, the number of ground nests increases.

The Eurasian Merlin chicks on the photo opposite have now almost fledged. *Vladimir Ivanovski.*

Morozov and co-workers note that in the northern taiga, although ground nests are known, Merlins almost invariably utilise stick nests, again chiefly of corvids, but also of Golden Eagles. In eastern Siberia, where the number of stick nests reduces, ground nesting is more common, Morozov *et al.* noting that in the Lena River valley ground nests were placed on a dry mossy bank, and in clumps of Rosemary (*Rosmarinus officinalis*) and Crowberry with occasional shrubby alder (Alnus spp). In one case the female falcon had laid in the notch where a branch left the stem of a larch (Larix spp).

Although an assumption might be made that the tree species favoured by Merlins for their chosen nests would reflect the preferences of the original nest builders, that was actually not the case in the study of Morozov and co-workers, Merlins showing a marked preference for stick nests in spruce, and avoiding nests in birch. The tighter crowns of spruces offer better protection than the open crowns of birches and so nests in the latter were only used if the falcons were avoiding ground nesting. In eastern Siberia where there are fewer spruces, larch is the preferred tree. In the southern taiga the same mix of stick and ground nests is used, the preferred tree species for nests in this area being pine or poplar. Some stick nests utilised in the west of the southern taiga zone are those of Short-toed Eagle (*Circaetus gallicus*) and European Nutcracker (*Nucifraga caryocatactes*).

In forest-steppe and steppe areas many nests are on the ground, usually sheltered by shrubs or tussocks, but Merlins will make use of any available cover, nests having been found in the houses and rubbish dumps of abandoned villages and in old haystacks. Stick nests are again mainly those of corvids (Magpies and Hooded Crows), often in pines, but also include those of the Long-legged Buzzard (*Buteo rufinus*) a bird of the semi-desert which fringes the southern edge of the steppe. In their studies of nesting throughout the Asian range of the Merlin Morozov *et al.* found no preference for stick nests at a particular height above ground, with nests found at heights from 4-21m.

Karyakin and Nikolenko (2009) studying three sub-species of Merlin in the Altai-Sayan area of Russia found that 87.1% of utilised nests were the stick nests of crows (Carrion Crows, Hooded Crows and Magpies), primarily in Siberian Larch. Other stick nests were of Rough-legged Buzzard and one of a Black Kite (*Milvus migrans*). The heights of nests varied from only 1.5m above the ground in willow to 28m in pines, though most nests were placed at 5-9m. Only one ground nest was found, that of an *F.c. lymani* pair.

While other bird species may occasionally choose exotic or unlikely nest sites, Merlins seem very conservative in their choices. However, they have been known to nest in artificially constructed sites. The usual constructions which find favour with the falcons are baskets, but more exotic structures have been used (see photos on the previous pages). But if Merlins nesting in an old bucket seems extraordinary, it has nothing on the tale told by Roderick McFarlane and recounted in Bent (1938). Writing in 1891 McFarlane states that Merlins were found along the Anderson River almost to Liverpool Bay in Arctic Canada and that one day 'a trusty Indian in my employ' found a nest with two eggs in a loose stick nest only six feet above the ground in the crook of a pine tree. The Indian shot at, but missed, the two parent birds. A few days later there were still two eggs, and again the parents were shot at, and missed. Days later the nest was revisited and was empty, but the parents were now nesting on a bank some 40 yards away. They had, says McFarlane, moved the original two eggs and added a third!

Breeding Density

In Saskatoon (Sodhi *et al.* 1992, in a study of urban-dwelling Merlins) found that the nearest neighbours of breeding falcons were at distances ranging from 161m to 4669m. The shorter of these distances implies very close nesting in comparison to studies in Britain where Newton *et al.* (1978), in Northumberland, measured average inter-nest distances of 1.0-1.6km, though with some up to 4.8km apart which concurs with the furthest distance measured at Saskatoon. Newton *et al.* noted that in the longer nest separations they observed, the intervening land between the nest sites had no suitable alternative sites, so it appears that breeding density is influenced by nest site availability as well as by the availability of prey. In his study of Merlins in England's south-east Yorkshire, Wright (2005) measured the closest separation at 350m, with the furthest separation 4.74km with a mean of 1.75±1.07km.

Wright (2005) also measured the orientation of the nests in his study areas. Orientation in this context was defined by the most open aspect of a nest placed on a slope. In several of the study areas the local topography meant that most sites were orientated to the south. However, in an area in which the topography allowed the Merlins a choice of either a north or south facing site, the chosen sites showed an equal number in each direction. This strongly suggests that the bird chooses a nest site with reference only to the openness of the view from it, rather than for any reason based on local wind direction or the position of the rising or setting sun. Wright's finding contradicts that of Sieg and Becker (1990) – see *Nest Sites* above – which found most sites were south-facing but, of course, as Wright suggests that choice is made on the basis of openness of aspect it may be that the topography in the Sieg and Becker study favoured southern aspects.

In studies carried out in Canada, a breeding density of 20 pairs/100km^2 was measured in Alberta (Smith 1978), with a density of 25.4 pairs/100km^2 being measured in Saskatoon (Sodhi *et al.* 1992). The latter is the highest breeding density recorded for Merlins, but relates to an urban population which represents a somewhat artificial environment for the species. Interestingly, the study of Sodhi and co-workers implied that the breeding density was self-limiting. The number of breeding pairs in Saskatoon had risen from 1 in 1971 to 31 in 1989, initially rising exponentially (Oliphant and Haug 1985). Plotting the population growth against the number of breeding pairs Sodhi *et al.* (1992) showed that growth (in terms of percentage increase) fell sharply as the number of pairs increased from 1 to 10, reaching about 10%/year, then decreased more slowly as the number of pairs increased further, finally reaching 0%/year (i.e. a stable population) when 30 pairs were breeding. Saskatoon, it seems, can accommodate 30 pairs of Merlin, but no more. The likely reason for this limit is the availability of prey as the nest spacings recorded by Sodhi *et al.* (1992) do not suggest that nest availability in Saskatoon is a limiting factor.

In contrast to the high breeding density of urban-dwelling Merlins, Becker and Sieg (1985) recorded a density of 3.8 pairs/100km^2 in south-eastern Montana, a value which is closer to that measured in *F.c. aesalon* in the two British study areas mentioned above: in Northumberland, Newton *et al.* (1978) calculated densities of 3-13 pairs/100km^2 over the years of study, with a mean of about 8 pairs/100km^2, while the average breeding density measured across the study areas of Wright (2005) over the 18 year period 1983-2002 (but excluding 2001 when access to the areas was prohibited because of an outbreak of foot-and-mouth disease) was 10 pairs/100km^2. However, the two areas covered by the studies of Newton *et al.* and Wright were 'prime' Merlin habitat. Another 'prime' Merlin area is the island of Lewis in the Hebrides, off Scotland's west coast, where there are no terrestrial predators to worry the ground-nesting falcons. In a study of the Merlins there in 2003 and 2005 Rae (2010) found a breeding density of 6 pairs/100km^2, the nest sites being spaced at a mean distance of 2.4km (range 1.4-5.8km). At a 'prime' site in Wales (moorland in the north-east of the Principality) Roberts and Jones (1999) found a density of 5 pairs/100km^2 over the period 1983-1997.

In a study of Britain as a whole Bibby and Nattrass (1986) concluded that a breeding density as high as 5-10 pairs/100km^2 was seen in areas of suitable habitat, but these were rarely extensive so that overall densities were lower: as an example they quoted a value of 1.7-2.2 pairs/100km^2 for the central Highlands of Scotland, a figure which accords well with that calculated for the moorland of the north of England as a whole (Nattrass *et al.* 1993), of 2.3 pairs/100km^2. In a study across Ireland (both Northern Ireland and Eire) covering the period 1986-1992, Norriss *et al.* (2010) found breeding densities varying from 1.22-5.89 pairs/100km^2.

The lower figures for northern Britain are consistent with those obtained by Nielsen (1986) who found a breeding density of 2.1 pairs/100km^2 in his study area of 5200km^2 of north-east Iceland. They are also consistent with densities calculated in Scandinavia: in a study in northern Finland, Virkkala (1991) calculated a breeding density of 3 pairs/100km^2 (for comparison, the density of breeding Eurasian Sparrowhawks in the same study was 10 pairs/100km^2). A similar figure was also obtained in a study of the breeding density of all Finnish raptors by Solonen (1994). Solonen found a relationship between breeding density and latitude, suggesting that the density was higher in northerly areas. However, his study only covered the country as far north as the Arctic Circle, while the study area of Virkkala (1991) was north of the Circle, implying that the Merlin breeding density increases as the birds move further from the more agricultural (and more populated) south, but then declines in Arctic Finland. Across the border in Russian Karelia the breeding density of Merlins appears to decrease, though this potentially represents the relative difficulty of survey work rather than a true decline. Zimin *et al.* (2005) quote values of 0.3-0.5 pairs/100km^2 in the forested areas of Karelia, rising to 1-2 pairs/100km^2 in the lower mountains of the Kola Peninsula. However, Zimin and co-workers note that in areas set aside as Nature Reserves or National Parks the density rises as high as 4 pairs/100km^2.

While the overall Merlin breeding density in Europe is low, as in North America there are habitat pockets where the density can rise significantly: in their study of Merlins in two Swedish National Parks (Padjelanta and Stora Sjöfallet), Wiklund and Larsson (1994) found an area of 12km x 5km in a valley which held 15 breeding pairs, implying a breeding density of 25 pairs/100km^2 suggesting a habitat that was extremely rich in prey resource: considering the entire area of the Parks Wiklund (1995) found an average breeding density of 4.75 pairs/100km^2. This is consistent with an estimate of 5 pairs/100km^2 in mountain birch forests and 2 pairs/100km^2 in the boggy areas of northern Sweden (Ottosson *et al.* 2012). It is likely that similar breeding densities occur in Norway and Finland.

Shubin (1984) calculated the breeding density of Merlins in the various biomes of the far east of Russia. He found that in the middle taiga (the denser part of the boreal forest) the breeding density was 0.53-1.58 pairs/100km^2, while in the northern taiga the density was 0.63 pairs/100km^2 (though in each case the sample size was small). In the forest-tundra, the section of taiga where the forest is thinning, the density was 2.7-7.5 pairs/100km^2, while on the tundra it was 2.5 pairs/100km^2, though again the sample size was small. From the number of nests found, the forest-tundra area was the preferred Merlin habitat. On the Bol'shezemelskaya tundra that links the northern Ural Mountain ridge (the Polar Urals) to the Arctic Ocean, Morozov (1997) found densities of 9-12 pairs/100km^2 in various years of study between 1982 and 1993. For further data on the breeding density of Merlins across Siberia see *Chapter 12*.

In their study in the Altai-Sayan region of Russia, where the three sub-species *F.c. aesalon*, *F.c. lymani* and *F.c. pallidus* breed, Karyakin and

Nikolenko (2009) found breeding densities of 1.0±0.6 pairs/100km^2, 5.0±2.0 pairs/100km^2, and 2.3±0.2 pairs/100km^2 respectively.

Copulation

Copulation often follows a food pass and may be initiated by either bird. The male may show his enthusiasm by staring at the female and, once he has her attention, bowing, fanning his tail and making the '*chip*' call. Alternatively, the female may show her readiness to copulate by bowing and fanning her tail to the male. Feldsine and Oliphant (1985) note that the male may also perch close to the female and make the '*kee-kee-kee*' call, or a single '*chip*' and that during a bout of staring at the female may make the '*chrrr*' (police whistle!) call, then 'flutter fly' towards her.

In one British pair of Merlins Stubbert (1943) noted that the two birds chased each other up and down a tree branch for 10 minutes, each moving sideways with down-spread tails, occasionally jumping up and fluttering on partly raised wings ('in the manner of cocks fighting') before eventually copulating.

Once copulation has been agreed, the female bends forward, fans her wings and raises her tail, moving it to one side as the male mounts her. The male beats his wings rapidly and fans his tail to maintain position and usually gives the '*kee-kee-kee*' call, the female responding with '*chip*' calls. Copulation takes 5-10 seconds and is repeated throughout the breeding season, Sodhi (1991a) noting about 60 copulations in total, the frequency increasing during the time prior to the laying of the first egg and then during egg production when the female is most fertile. Sodhi considers this high rate supports the 'social bond hypothesis' (**Box 20**). However, Sodhi also notes that as male Merlins cannot guard their mates at all times, high rates of copulation are also consistent with the 'sperm competition hypothesis' (See **Box 20** – the hypothesis is also considered again below). There is also a third hypothesis, the 'copulation trading hypothesis' (**Box**

Box 20

The three main ideas are, firstly, the social bond hypothesis, which contends that copulation is a way of establishing and maintaining the bond between a male-female pair, and so may be seen at times when the female is not fertile; the copulation trading hypothesis suggests that the female trades copulations for food, the male achieving breeding success by feeding the female sufficiently to ensure she does not seek food in exchange for copulation elsewhere; the third idea, the sperm competition hypothesis, posits that as the male is often away hunting it is in his interests to copulate often when she is fertile to maximise his chances of being the father of any resulting chicks.

20). In Sodhi's study only eight of 41 copulations (19.5%) occurred within 30 minutes of a food pass, and only three of the eight were initiated by the female. There is, therefore, no evidence in the Sodhi study that female Merlins exchange copulation for food. This was confirmed in a study by Dickson (1994) of Merlins in south-west Scotland. Dickson noted only three of 22 copulations (13.6%) occurring after a food pass. Dickson did note that 10 of the 22 copulations (45.5%) occurred before or after a nest visit by either sex.

In Alaska, Laing (1985) noted 11 copulation attempts by a male in 16.3 hours of observation, though not all of these were successful: there were three successful attempts in one 2.5 hour period. While copulation may continue throughout the breeding season, in established pairs, i.e. pairs which have wintered together or have migrated early in the spring, copulation may start significantly before egg laying commences, or even before nest selection has been made. It may even occur during the winter (Warkentin and Oliphant (1990) reporting copulation when the daytime temperature was a far from balmy -15°C). Such early copulation is considered to form part of pair bonding, and Warkentin and Oliphant note that in addition to copulation they witnessed food transfers and vocalisations in the normal winter-silent falcons, reinforcing the view that the birds were performing bonding behaviours.

After copulation the birds observed by Stubbert (1943) flew up, with wings almost touching, to a height of 150-180m, then turned simultaneously and swooped down, one above the other at great speed to tree-top level before turning upwards again and repeating the performance: the total post-copulation display lasted about an hour. However, while Stubbert's display must have been a joy to behold, most observers must be content with the birds enjoying a post-coital preening session.

While, as noted previously, Merlins are essentially monogamous, extra-pair copulations have also been reported. James and Oliphant (1986) report particularly interesting behaviour in urban-dwelling *F.c. richardsonii* in Saskatoon where there were several observations of a third bird aiding nest defence and food provision. In one case a first year male was observed taking and consuming food from an adult male, and also taking food which he then passed on to the female at the nest. The younger bird accompanied the adult male on hunting trips but was never seen hunting himself. The younger male also defended the nest, attacking both American Crows and Black-billed Magpies, when the adult male was absent. Similar behaviour has also been seen in Britain (Rebecca *et al.* 1988) (**Box 21** overleaf). Newton (1979) also records instances of male birds being replaced during the breeding season when the first male was shot.

In observations of other three-bird groups, the female was seen to copulate with the third bird. James and Oliphant speculate that the arrangement of a first-year male aiding an adult pair might be beneficial to the younger male, allowing him to learn behaviour which he could put into practice the following year, or to aid his own parents. However, as first-year males are capable of breeding, if copulation occurs (James and Oliphant saw this much less often than nest defence and food provision) the younger male may actually benefit much more directly: in such situations while it is clear that the female Merlin may benefit from enhanced nest protection and more secure food provision, it is unclear what benefit the adult male experiences.

Extra-pair copulations, or attempted copulations, were also observed by Sodhi (1991a) in the urban-dwelling Merlins of Saskatoon. Sodhi was attempting to gather information to differentiate between three main hypotheses put forward to define the copulation behaviour of birds (see **Box 20**). Sodhi noted that males were more likely to chase intruders away when their females were fertile, in keeping with the idea that the possibility of injury in a conflict when the female is not fertile is an unnecessary risk, but found that the observations were rather more ambiguous when attempting to differentiate between the copulation hypotheses. A later study by Warkentin *et al.* (1994) suggested the 'sperm competition hypothesis' was the more likely with respect to Merlins. Warkentin and co-workers took DNA samples from nine complete putative Merlin families (both putative parents and 31 nestlings) and nine other putative families in which only one putative parent was sampled, as well as 29 nestlings. The DNA fingerprinting showed no evidence of extra-pair fertilisation, nor any evidence of intraspecific brood parasitism, i.e. the dumping of eggs by one female Merlin into the nest of another female. While extra-pair copulations amounted to 7% of all copulations observed in various studies of the urban-dwelling Merlins of Saskatoon, the results of the DNA study of Warkentin and co-workers suggests that sperm competition results in male Merlins being responsible for the parentage of all, or the majority of, chicks resulting from their pair bonds. In a study in north-east Scotland using DNA profiling Marsden (2002) also found no evidence of polygyny.

Box 21
While it is widely assumed that third party birds are yearlings gathering experience, and observations of the behaviour of these birds suggests this is the most likely scenario, it is worth noting that it is not the only explanation which has been put forward for the existence of 'third birds' in the population. As Hunt (1998) has pointed out, all populations comprise breeding pairs, juveniles/sub-adults (which are not yet ready to breed) and 'floaters' which may be capable of breeding, but do not hold territories for various reasons. It is, perhaps, not always easy to differentiate between third bird helpers and third bird floaters. In work on Purple Martins (*Progne subis*) Stutchbury (1991) noted that male floaters did not intrude on all occupied territories equally, but preferentially chose a small number of territories. While attempts to usurp these territories initially failed in conflicts with the male holding the territory, the floaters fought a war of attrition which was often rewarded with long term success. This is interesting, but does not seem to accord with the behaviour of floaters seen in falcon territories.

Another theory is that late dispersal of juveniles might explain their presence in the nest area in the following year, though this theory would only explain floaters in non-migratory species: Kimball et al. (2003) also note that in most cases floaters are not family members, so late dispersal cannot be the sole reason for their existence. However, as Kimball and co-workers note, since floaters are much more prevalent in raptors than in other species there must be some evolutionary process at work. More recently, Campioni et al. (2010) in a study of Eagle Owls (*Bubo bubo*) in Spain noted that the behaviour of territory holders and floaters differed, the former sitting on prominent posts to emphasise their territory ownership, the latter being more secretive. This suggested that female floaters might actually be using the ability to move through occupied territories without being subject to an aggressive response as a way of assessing them for future occupation. This is consistent with the finding of Bruinzeel and van de Pol (2004), who, in a study in Holland, removed one breeding bird of a pair of Oystercatchers (*Haematopus ostralegus*) and observed how the resulting vacancy was filled. In 80% of cases it was filled by an 'intruding floater', i.e. a non-paired bird which had been seen to spend time within the territory of a breeding pair. Bruinzeel and van de Pol found that intruding male floaters were more successful at filling vacancies than females. While the breeding behaviour of Oystercatchers and falcons differ, the Dutch study suggests that it is possible that both male and female floaters in falcon species are gathering not only breeding experience prior to their own first breeding, but also acquiring information which might be of value later. In males this might allow a decision on which territories to compete for, while for females it might indicate which males/territories would be favourable. For an excellent review of the position for a raptor (the Eurasian Sparrowhawk) see Newton and Rothery (2001).

Breeding Part 2

In this Chapter we consider the final phases of the breeding cycle, egg laying, incubation and chick growth.

Egg Laying

With such a large range, in terms of both circumpolar distribution and latitude, and therefore differences in climatic conditions, it is not surprising that the date of laying of the first egg differs across the globe, varying from mid/late April through to late May. However, in studies in south-east Yorkshire, England from 1983-2002, Wright (2005) found the date of the first egg varied between 21 April and 6 June (Fig. 16), a variation which virtually covers the entire global range of observed first egg dates (in the urban-breeding Merlins of Saskatoon, Canada, Sodhi *et al.* 1992 observed first eggs laid between 19 April and 29 May). In their study of Asian Merlins Morozov *et al.* (2013) noted first eggs from as early as 27 April to as late as 9 June, the latter date, not surprisingly, being from falcons at the northern edge of the species' range. In their study of three sub-species in the Altai-Sayan region of Russia, Karyakin and Nikolenko (2009) found that *F.c. pallidus* laid clutches earlier than the other two sub-species. In his study of Merlin in Iceland Nielsen (1986) found that the laying of the first egg was dependent on ambient temperature, being, in

Figure 16

Variation of laying date of first egg in clutches of a study area in south-east Yorkshire, England.

Redrawn from Wright (2005).

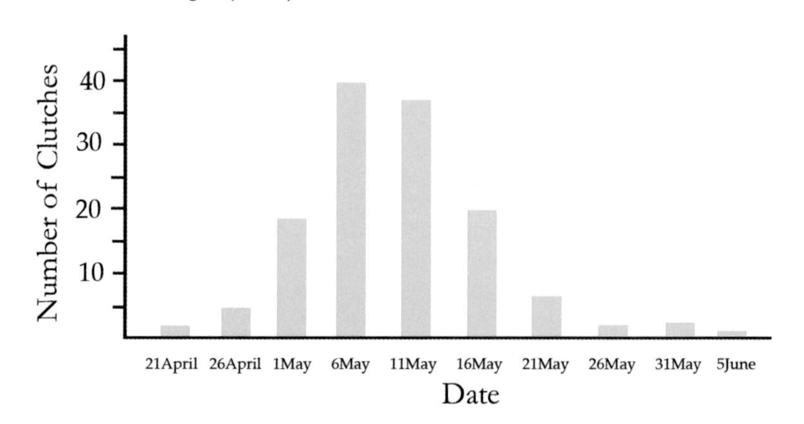

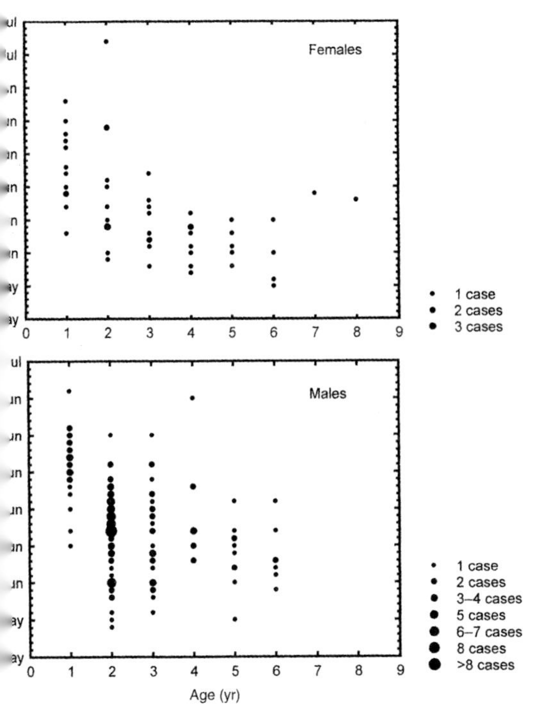

Figure 17
Scatter diagram of parent age and chick hatch date for male and female Merlins.
Reproduced from Espie *et al.* (2000).

general, earlier in warmer springs: in his study over a five year period Nielsen noted that the earliest first egg was laid on 9 May, the latest on 11 June. However, the earliest date (9 May) was actually noted in one of the colder springs, though the mean date of laying of the first egg in that year was later than in years with warmer springs.

The study of Espie *et al.* (2000) on the urban-dwelling Merlins of Saskatoon, found that hatching date was positively correlated with the age of both male and female Merlins (Fig. 17). The advantages of early egg laying, and, consequently, early hatching are clear. Merlins time their breeding so that juvenile prey species are abundant during the late nestling/fledging phase of their own cycle. For the Merlins, the vulnerability of the juvenile prey during this time, which are much easier to catch than their more experienced parents, allows the fledglings to benefit by gaining time to hone their hunting skills, both in terms of the possibility of easier prey, and in terms of maximising the number of 'training days' before the onset of winter. The parent birds also commence their moult at this time, so the possibility of easier hunting benefits them as well as their brood. However, the date of first egg laying cannot be so easily manipulated, as many studies of Eurasian Kestrels (by the Dutch group working at the University of Groningen and researchers in Fennoscandia) have indicated. Both the availability of prey and the influence of the increase in daylight hours on the female reproductive cycle affect the date of the first egg, though what is clear from the various studies, some of which have involved clutch manipulation, both reductions and increases in the number of eggs, is that birds of 'better quality' have earlier broods. In this context, 'better quality' means better hunters who have, consequently, survived the winter in superior condition and are able to take good advantage of spring prey, and mates with breeding experience.

Once established, the Merlin pair divides the workload of breeding between them in the standard format for northern hemisphere falcons. For a period before the female commences laying, the male feeds her, allowing her to build the resources necessary for the production of a clutch of eggs. The male continues to feed his female during egg laying and incubation, and for the first days after the chicks have hatched. The male also brings food for the chicks during this time. Sodhi *et al.* (1992) tabulated the division of labour in Merlin pairs nesting in urban Saskatoon (Table 19).

Phase of Breeding	Time fraction of Male (%)	Time fraction of Female (%)
Deliver Food		
During incubation	100	0
During nestling period	96.6	3.4
During fledgling period	65.0	35.0
Feed chicks	Only when female absent	Almost always
Incubation	7.2	91.8[1]
Brooding	0	100
Nest defence	57.6	32.9[2]

Table 19 Division of parental care by male and female Merlins. From Sodhi *et al.* (1992).

Notes:
1. The eggs were left unattended for 1% of the time.
2. 9.5% of attacks were initiated by male and female simultaneously.

Eggs

Once egg laying has begun, eggs are laid, on average, every two days (Palmer 1988) – Ivanovski (2003) suggests 36-48 hours from his study of *F.c. aesalon* in Belarus – though the last egg may be laid up to seven days after the penultimate. Each egg is elliptical and measures 33–43mm along the major axis and 28–33mm along the minor axis (*F.c. aesalon*, but consistent with the data of Sodhi *et al.* (2005) which includes measurements from the three North American sub-species). Ivanovski (2003) notes one egg which measured 44.5mm in length, and that length is also noted for some North American eggs (Trimble 1975). In his study of eggs Schönwetter (1967) considered that there was no discernible difference in the size of Merlin eggs across both its geographical range and the various sub-species, an opinion consistent with the data of Morozov *et al.* (2013) though the latter do not quote sizes for the eggs of *F.c. pacificus.*

Using the formula $V=0.51LB^2$, where V is the volume of an egg, L its length and B its breadth (Hoyt 1979), Rebecca (2006) calculated a mean volume of $20.1\pm1.3cm^3$ for eggs in his study area (**Box 22**).

Box 22

Roberts and Jones (1999) also calculated the mean volume of Merlin eggs in Wales assuming $V= LB^2$ and arrived at a figure of $38.9\pm0.2cm^3$ for the mean (and standard error) across 145 eggs which is in very good agreement with the value of Rebecca (2006) allowing for Hoyt's correction. Roberts and Jones noted that the mean volume of the eggs had fallen from $40.2cm^3$ (using their formula) for the pre-1983 period to $38.2cm^3$ for the period post-1983. They found no difference in the chick yield of nests in which eggs had above average volume against nests with smaller eggs.

Ground nest of Eurasian Merlins, northern England.
Mike Price.

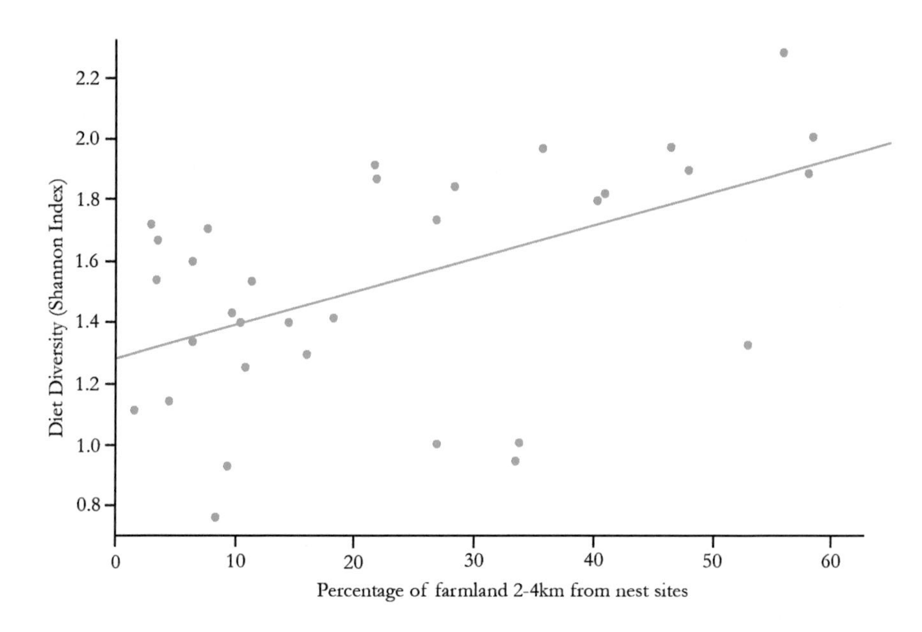

Figure 18 (*above*)
Variation of diversity of the Merlin diet in Wales with the percentage of farmland at 2-4km of the nest site.

Figure 19 (*below*)
Variation of volume of Merlin eggs in Wales with diversity of the falcon diet.

Both redrawn from Bibby (1987).

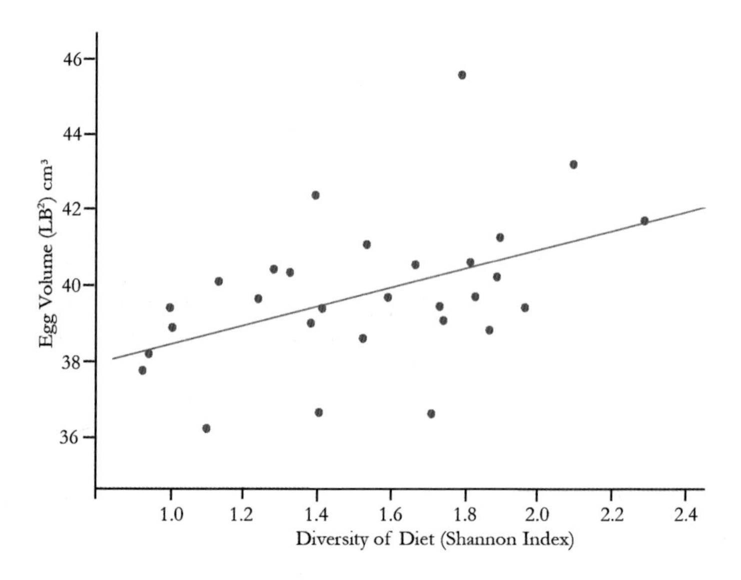

In a study in Wales Bibby (1987) also noted that the volume of Merlin eggs increased with the diversity of the diet of the falcons, i.e. the greater the number of species in the diet. Fig. 18 shows the variation of diversity in the prey taken with the percentage of farmland close to the Merlin nest site, while Fig. 19 indicates the variation of egg volume with diet diversity. While the variation in diversity in Fig. 18 is not surprising (farmland offering a habitat for a wider range of species) the effect on egg volume is much more so. But interestingly, egg volume was not correlated to breeding success, i.e. Merlin females which laid eggs with larger volumes neither produced larger clutches (though the sample size was probably too small for a correlation to have been identified) nor produced a greater number of fledglings. However, as the latter was significantly influenced by nest failures any correlation might have been masked.

Wright (2005) calculated a mean weight of 21.5g, with a range of 20-24g from a sample of 145 eggs in northern England. Wright also noted that the average egg weight at hatching was 17.9g (the egg losing weight as the chick develops). Again there seems no difference in the range of egg weights across the sub-species.

The base colour of the eggs is pale buff, this overlaid with blotches and spots of dark red-brown. However, there is a spread of egg coloration, with some having few dark markings, others having a base colour which is richer, this being occasionally due to heavy spotting of dark ochre.

Clutch Size

Clutches are normally 3 to 5 eggs, with a range of 1 to 7: Sodhi *et al.* (1992) record a clutch of eight in their study of the urban-breeding Merlins of Saskatoon but there was a possibility that the eggs had been laid by two females. Table 20 (overleaf) indicates the variation of clutch sizes in studies of European Merlins.

The mean clutches in Table 20 are consistent with the data of Nattrass *et al.* (1993) who calculated a mean of 4.2, with a range of 3.7-4.4 when studying data from seven regions across England. Rebecca (2011) found a mean of 4.2 and 4.3 in successive years (1993 and 1994) for clutches in nests across Britain.

The mean clutch sizes of Table 20 are also consistent with data collected in North America, where Sodhi *et al.* (1992) calculated a mean clutch of 4.2±0.04 from a sample of 237 clutches of urban-dwelling Merlins in Saskatoon, Canada, and also with the mean of 4.3±0.8 derived from a sample of 48 clutches in south-eastern Montana (Becker and Sieg 1985). They are also consistent with the data for Eurasian Merlins of Morozov *et al.* (2013) who note means of 4.3±0.3 on the Kola Peninsula, 4.1±0.2 on the Putorana Plateau, 3.9±0.6 and 4.5±0.1 in different studies of tundra clutches, 3.7±0.1 in the polar Urals, 4.8±0.2 on the Yamal Peninsula and

Percentage of Clutches with Various Numbers of Eggs								Study[2]
1	2	3	4	5	6	7[1]	Mean[3]	
0	2.1	9.8	52.1	35.1	1.0	0	4.2	Newton *et al.* (1986)
0	0.7	11.8	53.5	34.0	0	0	4.21	Wright (2005)
0	0	14.9	51.1	31.9	2.1	0	4.2	Nielsen (1986)
0	3	11	52	31	3	0	4.2	Ivanovski (2003)
2	0	6	59	30	3	0	4.25	Hagen (1952)
2	1	20	41	35	1	0	3.96	Brown (1976)
1.4	0	8.3	54.2	36.1	0	0	4.24	Roberts and Jones (1999)

Table 20 Variation of clutch sizes in *F.c. aesalon.*

Notes:

1. Seven eggs are rarely seen, but have been observed in Fennoscandia in 'peak lemming' years (e.g. Jourdain 1938). As noted above, Sodhi *et al.* (1992) record a clutch of eight.
2. Newton *et al.* (1986) sampled 194 clutches in Northumberland, England.
   Wright (2005) sampled 144 clutches in Yorkshire, England.
   Nielsen (1986) sampled 47 clutches in north-east Iceland.
   Ivanovski (2003) sampled 35 clutches in northern Belarus.
   Hagen (1952) sampled 63 clutches in Norway.
   Brown (1976) sampled 109 clutches across Britain. This work was carried out at a time when the Merlin population was recovering from the effects of organochlorine contamination, which may explain the lower mean in comparison to that of Hagen, which preceded use of the pesticides, and that of Newton *et al.* which was made substantially after usage had ceased.
   Roberts and Jones (1999) sampled 72 clutches in Wales. Their study covered 1964-1997 and so included both a period when organochlorine contamination was high, and the period of population that followed.
3. In a study of the variation of clutch size across Britain, Rebecca (2006) found variations from 3.2-4.9 in 1993 with a mean of 4.3 across 249 nests; and 3.7-4.9 with a mean of 4.2 across 259 nests in 1994.

3.8±0.4 on the steppes of Kazakhstan, though in all cases the sample sizes were small. The mean clutch sizes of Table 20 are also consistent with the data of Karyakin and Nikolenko (2009) who found the mean clutches of 3.8±0.9 eggs (range 2-5, N=17) the three sub-species of Merlin in their Altai-Sayan study area. However, their data was biased towards *F.c. lymani* (mean of 4.1±0.5 eggs, range 3-5, N=11) and *F.c. pallidus* (mean of 3.2±1.3 eggs, range 2-5, N=5) as only one nest of *F.c. aesalon* was sampled. The data of Morozov *et al.* (2013) suggests that there is no difference in mean clutch size between the Asian Merlin sub-species.

There are no recorded instances of Merlins breeding twice in the same season. Jourdain (1938), discussing the European Merlin sub-species, refers to 'second layings', but it is not entirely clear what is meant by this, though it is assumed by most experts that the reference is to replacement clutches. Jourdain says these are usually of three eggs. In the case of the study of Wiklund (1990a) on *F.c aesalon* in Sweden's Padjelanta National Park there is no ambiguity, the author definitely referring to replacement clutches. Wiklund notes that females laying replacement clutches must be early breeders, as late breeders do not have time to rear broods.

The data of Wiklund also show that there was no significant difference in the number of eggs in first and replacement clutches. This is consistent with a study of North American raptors and owls by Morrison and Walton (1980) who studied the literature and egg collections where the dates of collection were noted. Morrison and Walton noted that for most species clutch size decreased in replacements. However, for the Merlin the available data were limited, but implied that replacement clutches were of the same size. For most falcons replacement clutches are laid in the same nest, but the data for Merlins noted that some were laid 'nearby' (no definition supplied) or at distances of up to 300m. In Scotland, Cosnette (1984) noted a male mating with a second female after his first mate had been killed (probably by a mammalian predator which had also eaten the eggs). The second female nested less than 25m from the first nest and hatched three eggs.

In a study of urban-dwelling Merlins by Sodhi (1991a) instances of attempted intrusion to the nest of an established pair by another falcon were noted. In most cases the intruder bird was male, but two instances of female intrusion were seen. In both cases the intruder female went into the established pair nest, which implies an attempt at egg-dumping, but in each case the intruder was chased away by the resident pair before this (if it was indeed the intention) could be accomplished. In the case of the DNA fingerprinting of parents and nestlings by Warkentin *et al.* (1994) (see *Chapter 7: Copulation*) there was no evidence to suggest that such intraspecific brood parasitism had taken place, but the number of birds in the study was limited, so such parasitism cannot be ruled out.

Above The first egg of a Eurasian clutch at the point of hatch.

Below The hatching of this clutch of Eurasian Merlins is well advance.
Mike Price.

Incubation

The brood patches of adult birds develop prior to laying, feathers of the belly and lower chest area being lost to expose an area of richly vascularised skin, blood heat being transferred to the eggs. The stimulus for female birds to begin incubation may be visual or tactile. The tactile stimulus would be contact between the brood patch and an area/volume of eggs which would depend on final clutch size. In a study on Eurasian Kestrels (Beukeboom *et al.* 1988) it was concluded the eggs both stimulated incubation and clutch fixing by follicle suppression, though whether the stimulus was visual or tactile or both could not be differentiated. It is probable that the same mechanism operates to influence the decision of the female Merlin to begin incubation.

Although incubation is begun by the female the male Merlin also incubates, though in general contributes only a fraction of the total incubation time. In a study of urban Merlins in Saskatoon, Sodhi *et al.* 1992 observed males incubating for 7.2% of the total time (with females incubating for 91.8% of the time, the eggs being unattended for 1% of the time), while Laing (1985) estimated 15% of the time for Merlins in Alaska. Both these cases relate to nominate birds, with incubation fractions obtained by direct observation: in northern England in a study of *F.c. aesalon* Newton *et al.* (1978) considered the fraction to be higher at about 34%, obtaining this figure by observing which of the pair of Merlins was flushed from the nest on approach. In a later study Newton *et al.* (1986) flushed males on 65 of 230 nest visits (28.3%). A comparable percentage of incubation time (in this case 33%) was found by Temple (1972a) in a study in Newfoundland, also using the same 'which bird flushed' technique. In another study in northern England Wright says that females sit for periods of 2-4 hours and males for about half this time, and considered that females incubated for about 60% of the total time (i.e. males incubating for about 40%), in general agreement with both Newton *et al.* (1978) and Temple (1972a). Morozov *et al.* (2013) found that male incubation times varied from 8.7% to 35% in their studies. What is interesting in the incubation behaviour of Merlins, is how different it is from the Palearctic (Eurasian) Kestrel, a species in which the male does no (or very little) incubation, and in the Nearctic, where in the corresponding species, the American Kestrel, the male does incubate.

Rowan (1921-22 Part II) noted that incubation periods were 1-2 hours for each bird of the pair which would imply a higher fraction of total incubation by the male if he also shared night-time duties. Rowan noted one occasion when the male incubated through the night. However, as he points out, this was the only time he kept night-time watch on a Merlin nest and so was unable to decide whether this was standard behaviour or the idiosyncrasy of the individual bird. Most male incubation periods

Figure 20
Female Eurasian Merlin landing heavily among her eggs (*left*) and retrieving a displaced egg (*right*).
From Rowan (1921/22).

occur after he has delivered food to the female. When arriving with food the male makes the '*kee-kee-kee*' call and the females goes to him to receive the prey. She then flies to a convenient place, rarely out of sight of the nest and usually within about 150m, where she feeds, then preens. Occasionally the female will hunt for herself so that the male's turn at incubation lasts longer. Otherwise she calls to her mate and returns to the nest, perching nearby and waiting until he vacates so she can resume incubation.

The exact time of the start of incubation is still debated. Ruttledge (1985), who studied the breeding habits of captive *F.c. aesalon* pairs states that partial incubation begins once the third egg of a clutch of five had been laid and that continuous incubation begins only a day prior to the laying of the last egg. As most clutches appear to hatch synchronously, or nearly so (within one day – Palmer 1988) this seems the likely pattern. However, asynchronous hatching has been recorded which implies that some females begin full, rather than partial, incubation early.

During incubation the female will periodically turn the eggs: she will also give a soft call to initiate a response, re-assuring herself that her mate is present. Rowan (1921-22 Part II) noted that the female's habit of tucking her feet under the clutch meant that on occasions when she left a ground nest one egg would be disturbed a small distance from the clutch and she would recover it when she returned. On one occasion an egg rolled 30cm and resisted attempts at return as it had become trapped in its new position. After several attempts the female gave up and began to incubate the single egg, and then the remaining three eggs of the clutch in turn. Finally she made one further attempt to dislodge the single egg and successfully managed to return it to the clutch (Fig. 20). Rowan also noted that it was fortunate that the egg shells were strong as one female had the habit of landing on the clutch (Fig. 20).

Because of the difficulty of deciding precisely when incubation starts, the exact duration is not known, but lies within the range 28-32 days (Cramp and Simmons 1980). Mention must also be made of the extraordinary record of Ingram (1920) who found, in south Wales, a female Merlin sitting on three Grey Partridge eggs which had presumably been used as a replacement when her own eggs had been stolen. Ingram did not note the outcome of this situation but wondered, as will all others on hearing the information, what happened when the partridge chicks hatched – were they predated when, being precocial and essentially self-feeding, they left the nest, or did the female Merlin attempt to brood them and raise them on a diet of fresh meat.

Chick Growth

As with all birds, the hatchling escapes from the shell by chipping away at it with the egg tooth, the tooth being shed a few days later. Hatching takes about 24 hours, the emerging chick being semi-altricial and nidicolous. The chick weighs an average of 16.3g (range 13.0-17.5g) (Wright (2005) for chicks of *F.c. aesalon* in England: Morozov *et al.* (2013) quote comparable data) and is exhausted by the process of escaping the shell. The chick is covered with sparse down (buff and white on the upper parts, white on the underside) which dries quickly once the nestling is brooded by the female. Down that appears later, at between 4-8 days, is longer and more complete, and is pale brown/grey on the upper body, pale grey and white on the under body. The cere and feet are yellow, the bill blue and the egg tooth white. The chick's eyes are closed, opening after a few days. The female removes and discards, or, occasionally, eats, the egg shell.

Though essentially helpless, the chick can raise its head and will call softly for food. At first the nestlings are unable to thermo-regulate, i.e. cannot regulate or maintain their own body temperature, and so chill quickly if they are not brooded. This period lasts about 7 days (Sodhi *et al.* 1992), though studies in Britain suggest it may be as long as 10 days, and, in general, during this time the female is in constant attendance. Morozov *et al.* (2013) state that the period of intense brooding is longer for Merlins in the extreme north of their Asian range, noting brooding for 73-84% of the time for chicks 11-13 days old and 67% of the time at 14 days old. Not until the chicks were 21 days old did brooding cease.

During the period of intense brooding if the male brings food to the nest rather than landing nearby and calling to the female, she may be aggressive towards him, forcing him away (Temple, 1972c, quoting observations in Newfoundland). Wright (2005) agrees with this behaviour,

Recently hatched clutch in a ground nest of Eurasian Merlins, northern England. *Mike Price.*

but notes that it is not seen in all pairs, having seen male Merlins both feed and brood young chicks, though the brooding of young chicks was seen on only one occasion. This is consistent with the observations of Newton *et al.* (1986) who noted males incubating on only three occasions in 80 nest visits during the early stages of chick growth. The three occasions were when the chicks were one, five and ten days old, i.e. during the period when they were unable to thermoregulate, or shortly after they had become able to. It is not clear if the observed difference in these two studies represents a difference in the behaviour of sub-species or idiosyncratic behaviour by individuals or pairs. Once the chicks are able to control their own temperature (assuming, of course, that they are fed sufficiently often) brooding is not full-time. Rowan (1921-22 Part III) maintains that at the nest he observed the nestlings were not brooded at all after the first, critical, period. He writes, in a wonderfully evocative passage which says as much about the weather he was forced to endure as the chick rearing habits of Merlins, that 'from beginning to end I never saw the mother attempt to brood her young. No matter whether it thundered and poured in buckets, day and night they were left to themselves. At such times when quite young, they would huddle together, all four quite symmetrically arranged, breast to breast, four doleful little heads making a dome at the top of which the rain coursed down their backs in little runnels through the matted down.'

But it is not only the weather that makes Merlin huddle together in the nest. If a guarding adult gives an alarm call the chicks will immediately freeze in a huddle at the base of the nest so as to avoid aerial predators.

The nestlings are voracious and will make attempts to grab food from the female even when they are not old enough to cope with the prey themselves. Because of their enthusiasm for eating the chicks grow quickly, growth rates differing for males and females. Picozzi (1983) measured the increase in weight of 17 chicks in six broods on the Orkneys (Fig. 21a). In his study Picozzi noted that nestlings fell into two classes, those that attained a maximum weight of 225g, and those whose weight did not exceed 210g: he correctly assumed that the first class were females (nine chicks), the second were males (eight chicks). The variation in length of the outer (10th) primary with the weight of chicks on the Lammermuir Hills in south-eastern Scotland is shown in Fig. 21b (overleaf). The data, obtained from 154 male and 125 female Merlin chicks (for primary length) and 55 chicks for age, are consistent with that of Picozzi who measured the same parameter on the 17 Orkney chicks.

Figure 21a

Mean weight gain of eight male and nine female nestlings.

Drawn from data provided by Nick Picozzi. For full details see Picozzi (1983).

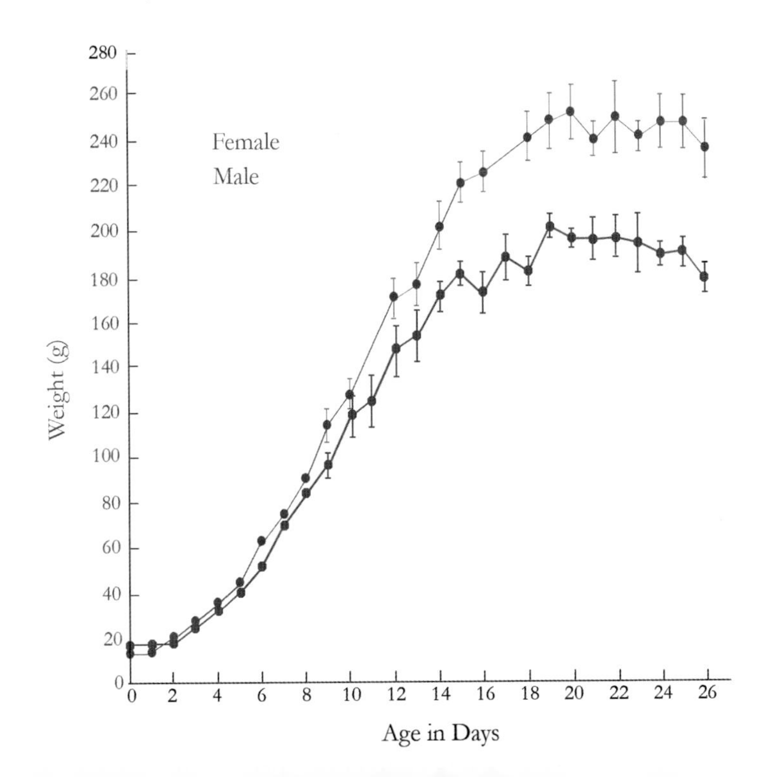

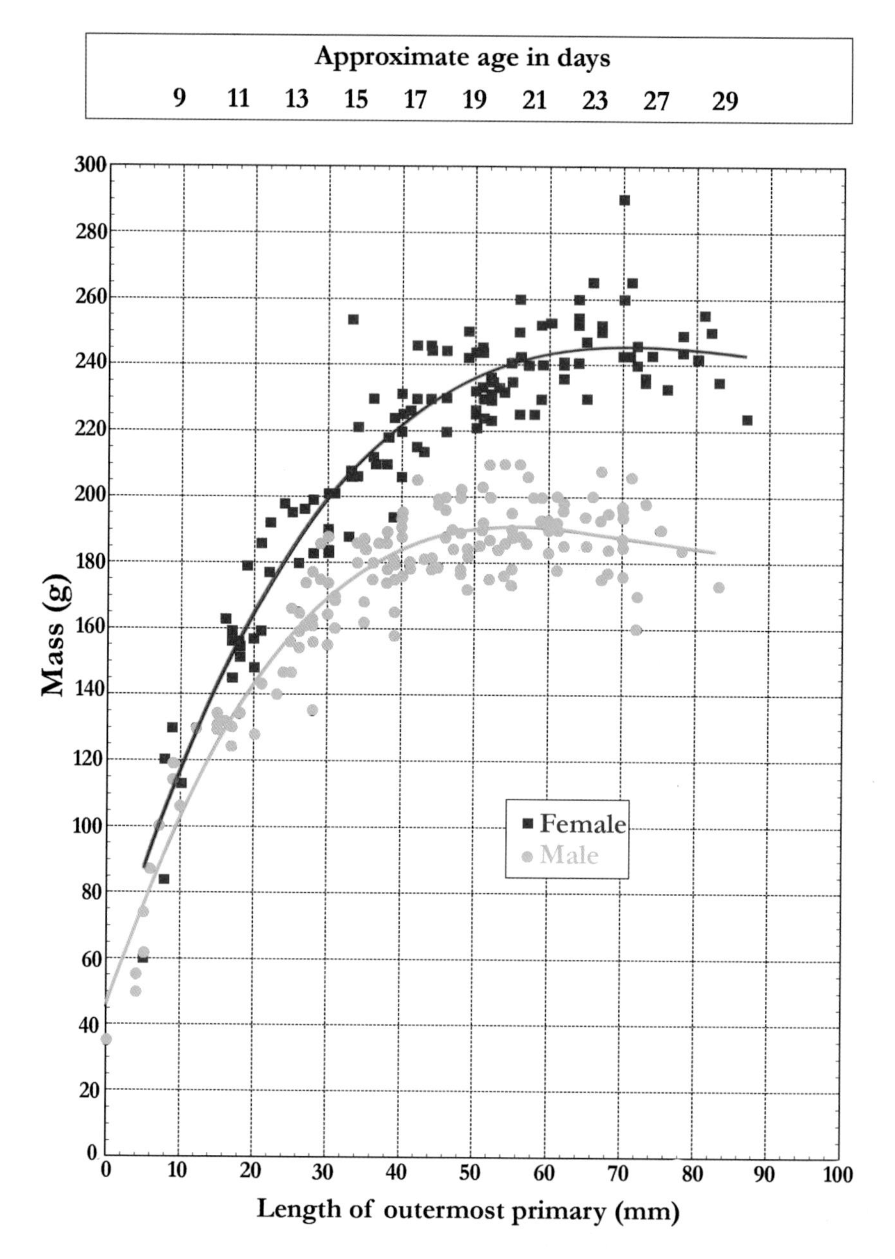

Figure 21b
Mean length of outer (10th) primary for male and female nestlings. The data on primary length was obtained from 154 male and 125 female nestlings from 70 broods over a 15 year period in the Southern Uplands of south-east Scotland. The age data were obtained from 55 nestlings of known age.
Drawn from data of Ian Poxton and Alan Heavisides, by permission of Alan Heavisides.

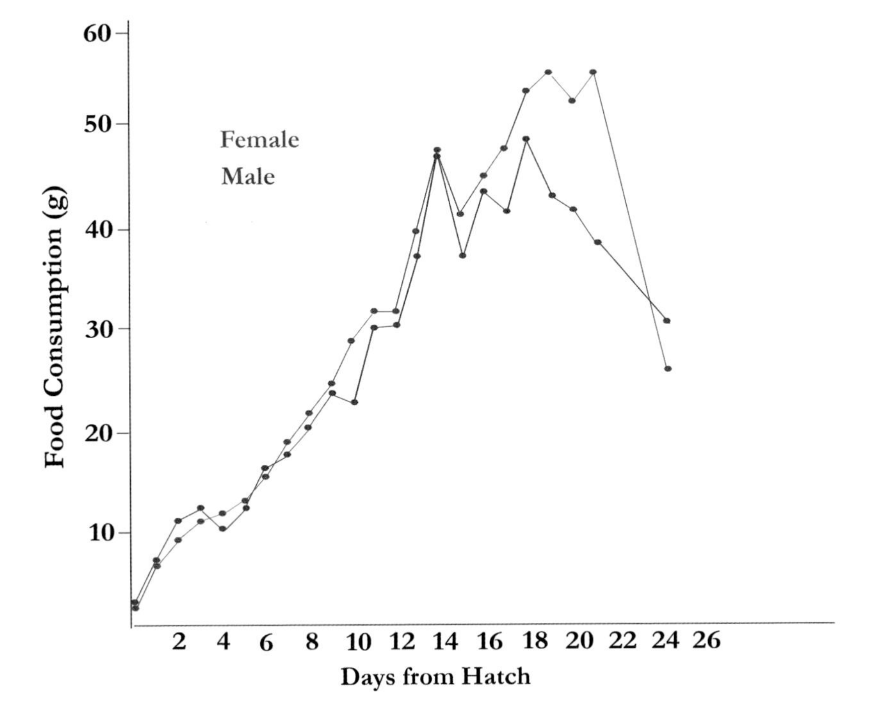

Figure 22
Food consumption in grams/day by one pair of male and female chicks as a function of time from hatching.
Redrawn from Oliphant *et al.* (1985).

The weight curve for the chicks follows the familiar sigmoid curve, and indicates that the period of maximum growth is between days 5-15 from hatching. The weight data is consistent with that of Wright (2005) on the same sub-species in northern England, and with studies on two pairs of captive North American *F.c. richardsonii* chicks (Oliphant and Tessaro 1985). One of the two pairs of chicks were taken from a wild nest about eight days after hatching, the other pair were taken from a wild clutch and reared in the laboratory from egg stage. In the latter case the weights at birth of the two newly-hatched chicks were 13.0 and 13.7g. Oliphant and Tessaro note that the final weights achieved by the chicks in their study were lower than those measured in wild Merlin chicks, and while the sample was too small to mean that the findings were in any way statistically significantit was considered that this finding was consistent with a problem noted in other captive-reared chicks. Oliphant and Tessaro consider the likely cause is dehydration as the food items fed to wild chicks include fluids added from the nasal glands of the parent birds. (As a consequence, Oliphant and Tessaro suggest the addition of water or physiological saline to food items for captive reared chicks would be prudent.)

Oliphant and Tessaro (1985) also measured the food intake of the two pairs of chicks. Fig. 22 indicates the food consumed by the pair hatched in the laboratory and shows an essentially linear increase from hatch to

Day 14 from hatch, with the male chick taking 40.0g/day and the female taking 46.1g/day. (The food intake graph for the chicks taken from the nest is similar, but indicates a more erratic intake.)

Food for the growing nestlings is provided by the male Merlin. In a study of Merlins in the Denali National Park the male delivered 3 birds/day to the female throughout the incubation period, the delivery rate rising to 9.6 birds/day during the nestling phase (average delivery to nests with an average of 3.75 chicks – Laing 1985). Morozov *et al.* (2013) quote provisioning rates of 11 times daily to three 7-day old chicks, and 14 times daily to a two chick brood at 14 days of age.

In general the male plucks the prey before delivery using a favoured plucking perch which often lies within 150-200m of the nest. Usually the head and wings of the prey will be removed, as well as the feathers. For ground nesting birds the plucking 'post' may also be an area of bare ground among the vegetation, but is more usually a nearby boulder, the boulder becomes conspicuously covered in whitewash from the bird's defecation as the season lengthens, as well as with discarded prey detritus. Usually the boulder will be in direct line of sight of the nest, often on the opposite side of the valley from which the nest is set. Boulders are also commonly used as plucking posts if the nest is in an isolated tree, but a nearby tree is used if the nest is within a wooded area.

Once the food is delivered to the female, she will initially break off small sections of the prey and feed each chick bill to bill, but soon the chicks are snatching at any available morsel and attempting to swallow it whole. Craighead and Craighead (1940) tell a delightful story of a female engaged in a tug-of-war with four nestlings which successfully pulled prey away from her only to discover they were completely unable to dismember it and had to wait until the female regained possession and had torn off pieces the youngsters could cope with. With the meal completed, one nestling grabbed the remains of the prey and mantled it, despite being unable to either swallow it or tear lumps from it. A sibling then grabbed it, with the same outcome. As Craighead and Craighead note – 'the remains of the (prey) went the rounds: none could swallow it and all were too young to tear it up'. Once they are old enough to manage their own food, taking any larger piece they are able to grab, the nestlings will mantle the food to avoid their siblings from taking it.

In the studies of urban-dwelling Merlins in Saskatoon, Oliphant (1974) did not observe the female hunting during the nestling phase of breeding, while Sodhi *et al.* (1992) saw the female hunt on only three occasions, each of which occurred after the male had failed to deliver food for three hours. These studies suggest that females do not help in the provisioning of the offspring until they have fledged. Lawrence (1949) reports observations of Merlins nesting at Pimisi Bay, near North Bay, Ontario. In

Box 23
Lawrence (1949) had some delightful observations about the Merlins, including a graphic description of a hunt in which one of the pair 'came like a bullet towards my window, but within three feet of the pane made a right-about turn' and scattered a flock of chickadees. Lawrence also noted the male predating Eastern Kingbird chicks at the nest, having arrived there after a failed attempt to take a parent bird. While the Merlin peered into the nest 'the kingbirds, oddly inefficient, screamed and flitted around the hawk which defended itself merely by raising its wings'. It is not clear what more the kingbirds could have done, which raises more sympathy for the kingbirds with the reader than Lawrence seems to have had: it appears the Merlin returned and took all the chicks. Ultimately the female Merlin also disappeared. Lawrence assumes both were shot, and to her credit she then rescued the two Merlin chicks, raised them to fledge and was rewarded by seeing them learning to hunt and, eventually, dispersing.

these the female did not hunt until the male Merlin disappeared (assumed shot) when the chicks were 11 days old (**Box 23**).

However, while all these studies imply females do not hunt until the young have fledged, in a study of Merlins in Newfoundland Temple (1972a) noted that the female started to hunt seven days after the eggs had hatched, and that her contribution to the nestling diet gradually increased during the nestling phase. While this study was on nominate Merlins it seems probable that the differences arise between individual birds rather than sub-species.

Debris at a favourite plucking place of a male Merlin preparing prey for delivery to the nest.

The study of Oliphant and Tessaro (1985) on captive-reared chicks allowed the development of the young Merlins to be observed (Table 21).

Days from Hatching	Status of development
1	Eyes closed, down wet, but dries quickly
3	Eyes open
4-8	Second down develops
9-11	Sheathed contour feathers appear Can sit erect on tarsi Wing flapping commences
12-14	Primaries break from sheaths First cast
15-17	Secondaries emerge Retrices break from sheaths Egg tooth lost Stands upright[1]
18-21	Head and body feathers develop Flight feathers develop rapidly Holds food with feet Wing flapping increases Weight gain and food consumption level off
22-28	Maximum weight attained Down replaced by contour feathers (down only visible on head at 28 days) Makes short jump flights
29-34	Fledges and makes short flights[2]
35-40	Long flights made Notices prey and makes first attempt at pursuit
40-50	Begins to hunt and kills first prey (usually insect) Feather growth complete
50-60	Becomes independent and disperses[3]

Table 21 Development of young Merlins, From Oliphant and Tessaro (1985) with minor amendments from other sources.

Notes:
1. While feather development moves quickly, the birds locomotive skills lag behind, nestling Merlins being unable to stand upright until they are

15-17 days old, and unable to hold food in their feet until they are 18-21 days old. Remarkably, the day after being able to successfully hold food in their feet some birds take their first flight, though some nestlings will be 28 days old before they are fully fledged. The young of ground nesting Merlins usually abandon the nest and disperse among the surrounding heather once they can walk easily, though they do not move far so as to be able to reach the female quickly if she arrives with food. The dispersal aids both camouflage and sanitary conditions at the nest, the chicks being unable to avoid nest spoiling by defecating over the nest edge as those in stick nests do.

2. Studies of *F.c. aesalon* state that fledging is complete at 28-32 days.

3. Once fledged, the juvenile birds stay in the vicinity of the nest and continue to be fed by their parents, which pass food either aerially or at perches. Aerial passes seem to be a method of improving the juvenile's flying skills as well as their hunting skills. The juveniles also hone their hunting skills by chasing feathers, leaves or thistle-down, and also by chasing insects, quickly developing the agility that is a characteristic of the species. The juveniles become independent of parental care 2-4 weeks after fledging.

Eurasian Merlin chick at 12 days of age.

In poor, and rapidly changing, light a drama plays out in a basket nest in Scotland.
Above Two chicks venture away from the nest.
Below One chick soon returns, aware that food is only delivered to the nest.

Above The fourth chick, which was having trouble getting back into the basket, looks on enviously as a sibling exercises his wings.

Below The fourth chick finally makes it back into the nest and the four actors take their bows.

Breeding Success

Studies across the Merlin range which have looked at the success of breeding pairs, in terms of the number of chicks raised to fledge, has shown the success rate to be variable. In their study of urban-breeding *F.c. richardsonii* in Saskatoon Sodhi *et al.* (1992) found the falcons were remarkably successful. In only 4% of nests (10 of 256 nests observed between 1971 and 1990) did no eggs hatch. In those nests that did produce hatchlings, an average of 4.2±0.04 eggs produced 3.8±0.03 chicks, a success rate of 90.5%. A comparable success rate (84%) was seen in a rural population of the same Merlin sub-species in south-eastern Montana (Becker and Sieg 1985). In the latter study a success rate in terms of eggs producing fledged young (76%) was calculated in addition to the number of eggs hatching (84%), indicating that if an egg hatched the hatchling had a 90.5% chance of fledging. Morozov *et al.* (2013) quote a figure of 7.4% for egg mortality based on a study in Belarus.

In the prairie lands of Alberta, in a study across four years (1971-1974) Hodson (1976) recorded a mean clutch size of 4.6, but a mean brood size of 3.5, with a mean of 3.2 eggs producing fledged young. These mean values convert to a success rate of 76% (eggs to brood), and a 91% chance of a hatchling surviving to fledge, comparable to the percentages found in a study in Alaska's Denali National Park. There, Laing (1985), observing only four nests, each of which held a clutch of five eggs, noted a hatching success rate of 75%, and a 93% success rate in terms of hatchlings reaching fledge. However, the full data from the study of Hodson (1976) reveal that the success rates as defined by mean clutch and brood sizes, though allowing comparison to the data of Laing (1985), overstate the actual productivity, and therefore the overall success of the breeding Merlins. Only 70.9% of occupied nest sites resulted in eggs being laid, and only 54.9% of nests with eggs hatched chicks, though 85.3% of these successful nests produced fledged birds. One storm caused the failure of over 41% of all the nests in one part of Hodson's study area.

At first glance the lower success rate of Hodson (1976) might be thought to be due to the effect of residual organochlorine contamination, though that is unlikely to have been the case for the study of Laing (1985). However, in a an earlier study by Fox (1971) it was noted that in eastern Canada the hatching success rate in Merlin prior to 1950 was 81%, whereas in the period 1950-1969 it dropped to 48%. In the Great Plains area the pre-1950 success rate (again measured as successful hatching) was 92%: in the period 1950-1969 this rate fell to 49% on the prairies, but rose

Opposite
Female Eurasian Merlin delivering food to her nestlings, Belarus.
Vladimir Ivanovski.

to 98% in forested areas where the Merlin prey species were much less likely to become contaminated. In northern Canada no success rate figure was available for the pre-1950 period, but for the period 1960-1969 it was 89%. It is therefore considered that the Hodson (1976) success rate data are probably not low as a consequence of residual organochlorine contamination but result from differences which can be seen across the range.

For other areas the breeding success rate can differ markedly from that seen in the best returns from North America. In Iceland (Nielsen 1986) during 1981, when springtime was particularly cold, only 71% of occupied nests produced young, this figure rising to 84% in years where the spring was more amenable, in terms of climate. The success rate of those nests which did produce young varied from 70% to 84% across the five years of study (1981-1985). In his study period Nielsen followed 42 clutches from laying to fledging. Of these, seven clutches (17%) failed completely: two were deserted; four disappeared, of which one was destroyed by humans and at least one other was taken by humans; and in one no eggs hatched. In successful nests 13 eggs failed to hatch, 14 eggs or hatchlings disappeared with no trace as to the reason and six nestlings died: of the latter, four died in one severe rain storm.

In the study of Newton *et al.* (1978) in northern England it was found that about 37% of nest areas which were occupied by Merlins at the start of the breeding season did not result in nests with eggs, i.e. that the breeding attempt of over one-third of apparently paired birds failed before the egg-laying stage, a failure rate at that stage comparable to that of Hodson (1976). Newton and co-workers noted that if breeding failed then both birds tended to leave the nest area within a few days. In their study of Merlins in Ireland, Norriss *et al.* (2010) also found comparable figures to those of Newton *et al.* (1978) with 33% of pairs deserting their nests before eggs were laid. Norriss also found that 90% of nest desertions occurred during the pre-hatching phase, the only exceptions being two broods of chicks less than 10 days old which failed during prolonged heavy rain: no mortality of older nestlings was observed. This finding, that once eggs have hatched Merlins are consistent and effective parents, is also confirmed by other studies.

Newton *et al.* (1978) also looked at the success rate, in terms of the number of eggs laid which resulted in hatched chicks, for those pairs which did lay, and identified the causes of egg failure in clutches which were entirely lost (Table 22).

As can be seen from Table 22 predation of eggs and nestlings is a major source of loss from Merlin nests and we return to this below.

od	Eggs not laid	Eggs broken by parent or predator[2]	Eggs taken	Eggs deserted	Eggs addled[1]	Eggs trampled	Nestlings taken by human or predator	Female died	Unknown[3]
1970	1	3	1	0	4	1	0	1	7
1973	4	6	2	0	1	0	5	0	2
1976	1	7	13	1	0	0	2	0	(1)

Table 22 Causes of complete nest failure in period 1961-1976 in northern England. From Newton *et al.* (1978).

Notes:

1. Egg addling, along with eggshell thinning, is a known effect of organochlorine contamination.
2. It was not possible to determine any significant differences as a result of a decline in eggshell thinning during the period because of the difficulty of assigning egg breakages to specific causes. However, as the number of nestling successfully fledged during the period increased and the number of instances of addled eggs declined, this implies that overall egg failures resulting from organochlorine contamination declined.
3. It is notable that Newton *et al.* (1978) do not specifically note the loss of clutches as a result of adverse weather (though 'unknown' may include this). In other studies the weather has been responsible for losses, e.g. Hodson (1976) who noted that in his Canadian prairie study area after a storm that lasted two days and which included winds to 80kph, over 10cm of rain and a fall of 17°C in ambient temperature, the number of active Merlin nests had fallen from 26 to 15, a 42% loss. Of the 11 inactive nests, one had two dead nestlings, six had cold, dead eggs, two were empty, and two had egg shell remains: for one of the latter nests the shells were found some 6m beyond the nest which had itself been blown 5m out of the tree in which it had stood. Hodson considers the shell remains were likely the result of crow predation. Of the still active nests, dead nestlings were found in two. Other species nesting in the same study area also suffered losses.

Converting the data from Table 22 into overall mortality, but only for the period 1971-1976 so as to eliminate, as far as possible, the effect of organochlorine contamination, Newton *et al.* (1978) were able to conclude that egg losses from all causes were responsible for the loss of 1.5 chicks per brood. Of a total of 78 nests in which it was known with certainty that losses at the egg stage from known causes had not caused brood depletion losses at the nestling stage were minimal (Table 23), i.e. if an egg hatched the chances of the chick reaching fledging were good.

	Number of Broods	**Mean number of nestlings lost[1]**
All nestlings fledged	59	0
Nestling mortality	6	1.2
Early loss[2]	13	1.0

Table 23 Brood losses at the nestling stage. From Newton *et al.* (1978).

Notes:
1. The data on nestling losses are consistent with the data of Roberts and Jones (1999) in north-east Wales where a mean clutch size of 4.24 eggs produced a mean of 3.1 fledglings in the period 1964-1974, 3.7 fledglings in 1975-1982, but then 2.9 fledglings from 1983-1997, and with that of Rebecca (2011) who found 3.1 fledglings from a mean clutch of 4.3 eggs in 1993 and 3.5 fledglings from a mean clutch of 4.2 eggs in 1994 from nests across Britain. The data are also consistent with that of Ivanovski (2003) for *F.c. aesalon* in northern Belarus who found that over the period 1991-1997 a mean clutch of 4.2±0.8 eggs produced 3.4±1.1 fledglings, and with Karyakin and Nikolenko (2009) in their study of three Merlin sub-species in the Altai-Sayan region of Russia. They found a mean clutch size of 3.8±0.9 eggs and a mean brood size of 3.2±1.0 (range 1-5, N=28) successful fledglings implying losses of the order of a single chick. The data of Karyakin and Nikolenko was biased towards *F.c. lymani* which provided 23 nests of their 28 samples.
2. It was unclear in these cases whether the loss was at the egg stage or when the chick was very young. However, hatching is exhausting for young birds and losses at that stage are known in comparable species. It is also the case that some hatchlings may suffer developmental abnormalities which may result in death at an early stage, or after fledging

Eurasian Merlins, Scotland.
Graham Rebecca.

when the juveniles are unable to fend for themselves. While rare, such cases are known. Cooper (1984) records a case of a three-week old male Merlin taken from the nest with the third and fourth digits of the right foot fused. The digits were separated under anaesthetic and the bird recovered, but then, sadly, died of Dieldrin poisoning.

The fate of young which die in the nest is unclear. Temple (2008) found all five nestlings from one nest had been plucked, four in/close to the nest, one at a plucking post about 100m away: the latter had been partially eaten. Temple believes the chicks had died following several days of continuous rain and had been eaten by the adults. Whether dead chicks are consumed by siblings in other instances, as is the case in some species, is not known.

Newton *et al.* (1986) continued the study of Merlins in Northumberland, England through to 1983 and were therefore able to update the information of nest failures and breeding success. At the time the population of Merlins was in decline in the area. Combining data from the earlier study with the new work to allow a ten-year period (1974-1983) there was a decline in observations of activity in potential nest areas from around 60% of sites in 1974 to about 25% in 1983, and a decline in nests found from about 38% in 1974 to about 20% in 1983. Newton *et al.* repeated their analysis of the causes of complete failure of nests over the ten-year period (Table 24 overleaf).

	Eggs not laid	Eggs broken by parent or predator	Eggs taken[1]	Eggs deserted	Collapse of tree	Nestlings taken by human or predator[2]	Female died
Total Failures	13[3]	21[4]	46	13[5]	2	25	2[6]
Percentage of Total Failures	10.7	17.2	37.7	10.7	1.6	20.5	1.6

Table 24 Causes of complete nest failure in 1974-1986. From Newton *et al.* (1986).

Notes:
1. Includes 6 definitely taken by humans.
2. Includes 7 taken by humans and 7 definitely taken by natural predator as remains were discovered.
3. Includes a female taken by a Peregrine Falcon in 1981.
4. Includes one nest trampled in 1982.
5. Includes two addled clutches.
6. Includes a female shot in 1978.

In total 370 sites with signs of activity were observed in April (of the study period of 1974-1983), but only 230 nests were found, so 38% of apparent breeding activity did not result in eggs being laid, in agreement with the earlier study (Newton *et al.* 1978). Of the nests found, 217 had eggs, and of these chicks were fledged in 143. What is clear from Table 24 is that predation of both eggs and nestlings is the most significant reason for failure, and in the following section we explore this aspect of Merlin breeding in more detail. In their study of Merlins in the Altai-Sayan region of Russia, Karyakin and Nikolenko (2009) found two instances in which nests had failed because chicks had been predated by a mustelid, and three in which death of the female due to predation (one by a mustelid, one by a Goshawk and one by a Saker Falcon) had precipitated nest failure.

From studies in North America (Sodhi *et al.* 1992), the earlier the Merlin pair bred the more likely they were to lay a larger clutch and to produce fledglings, clutches in which the first egg was laid in April yielding 4.4 fledglings from 4.7 eggs, compared to 3.8 fledglings from 4.3 eggs for nests where the first egg was laid in May. (The standard deviation in both cases is 0.1, and the differences in both egg and fledgling numbers are statistically significant at the 99% level.)

As already noted (see *Chapter 7: Pair Formation*) Espie *et al.* (2000) also found that the breeding success of Merlins improved with age. In females, this was due to higher mortality of less successful breeders rather than any specific change in individual birds (though, of course, the gathering of experience cannot be detrimental), while among males, changes in breeding performance with age were largely due to the experience gained by males in the first two or three years of life. In essence the results of Espie and co-workers mean that if a falcon of either sex survives from one year to the next its breeding performance will improve, until the natural deterioration of age forces a decline.

Nest Predation

The findings of Newton *et al.* (1978, 1986) that predation was the major cause of reproductive failure in Merlin was confirmed in several studies by Wiklund in Arctic Sweden. There, predation at the egg stage is primarily by Hooded Crows, while at the nestling stage is primarily by Rough-legged Buzzard and Weasel (*Mustela nivalis*). In one study Wiklund (1990a) noted that reproductive success was in large part due to the nest defence tactics of the breeding Merlin pair. While this result might seem obvious as there are a limited number of ways in which the falcons can protect against predation, choice of nest site and nest defence being the clearest options, the Wiklund study offered extremely interesting insights into the behaviour of breeding birds and the consequences of that behaviour. Wiklund notes that while the investment in reproduction is equal overall, in that if breeding is successful the sexes are equally rewarded in terms of surviving offspring, the investment is time dependent as it can be argued that females invest more in the early phases of breeding and therefore would benefit from mating with males who would contribute the same investment effort by offering greater nest defence when they are solely responsible for it, the female being essentially tied to the nest. There is some evidence to support this idea from other species where males mate with several females and so demonstrate their defensive qualities to females who are mated later, but it is not clear how females who are essentially monogamous make such a choice. The problem is further complicated by the fact that defence against a potential nest predator is risky as most will also be adult bird predators, so any defence must weigh risk against the benefits.

Wiklund (1990a) studied Merlins breeding in northern Sweden's Padjelanta National Park where the falcons are not threatened by local Reindeer (*Rangifer tarandus*) herders, so eliminating one source of 'predation' (i.e. by humans) which had been significant in the British studies. To study Merlin behaviour Wiklund placed a Raven (a taxidermy example) on a pole in open ground close to nest sites. The Raven was chosen as it is both a significant nest predator but also a formidable adversary, very capable of injuring or killing a Merlin. The stuffed Raven was presented to a total of 29 Merlin pairs over a two year period, during the egg laying phase and again while the chicks were newly hatched. The Raven was positioned for various times to study the effect of predator exposure times.

For 25 pairs, at least one Merlin attacked the stuffed Raven. Attacks involved dives which went within one metre but on occasions involved a strike to the stuffed bird's head or back. Male attacks were steep, fast stoops, while female stoops were slower and shallower. In the cases where no attacks were launched, the Merlins either left the nest area immediately, or circled above the Raven at about 5m making the alarm call, and then left the area. Wiklund found that during the egg laying phase males attack more frequently than females, and the frequency of attacks made by the male was positively correlated to clutch size, while the female attack frequency was negatively correlated with clutch size (Fig. 23a), the two frequencies effectively cancelling so that the overall number of attacks was not correlated to clutch size. Of the 29 male Merlins, 15 made attacks (varying in frequency from once to 20 times), 14 did not. Of the 29 females 14 made attacks (varying from once to 12 times), 15 did not. When the chicks had hatched, the female attack frequency increased, but there was no difference in the behaviour of males (Fig. 23b). Overall the number of attacks increased with brood size as a consequence of the positive correlation of female attacks with brood size. These findings suggest that male Merlins assess the value of nest defence by the size of the clutch, but it is the female, who does most, often all, of the feeding of chicks who is able to assess their value, in terms of fitness to survive.

Wiklund (1990a) found that a third of Merlin nests failed due to predation, and that failure of other nests might have been due to the presence of potential predators in the area which cause the female to desert, occasionally to lay replacement eggs in another nest. (One problem of the Merlins being dependent on the nests of Hooded Crows for breeding is that usable crow nests are, by definition, in areas favoured by the crows and the crows are potential egg and nestling predators.) From Wiklund's study it was clear that male nest defence was important in reducing the likelihood of desertion. In pairs in which the male attacked less than the female the frequency of desertion was high. Females heavy with eggs are less agile than males, so attacks on predators are more risky. In situations in which the male is not pulling his weight, as it were, the female has decisions to make: does she make attacks, and if so at what frequency? At a high frequency she risks injury or death, while at a lower frequency the likelihood of egg predation increases. Or does she abandon

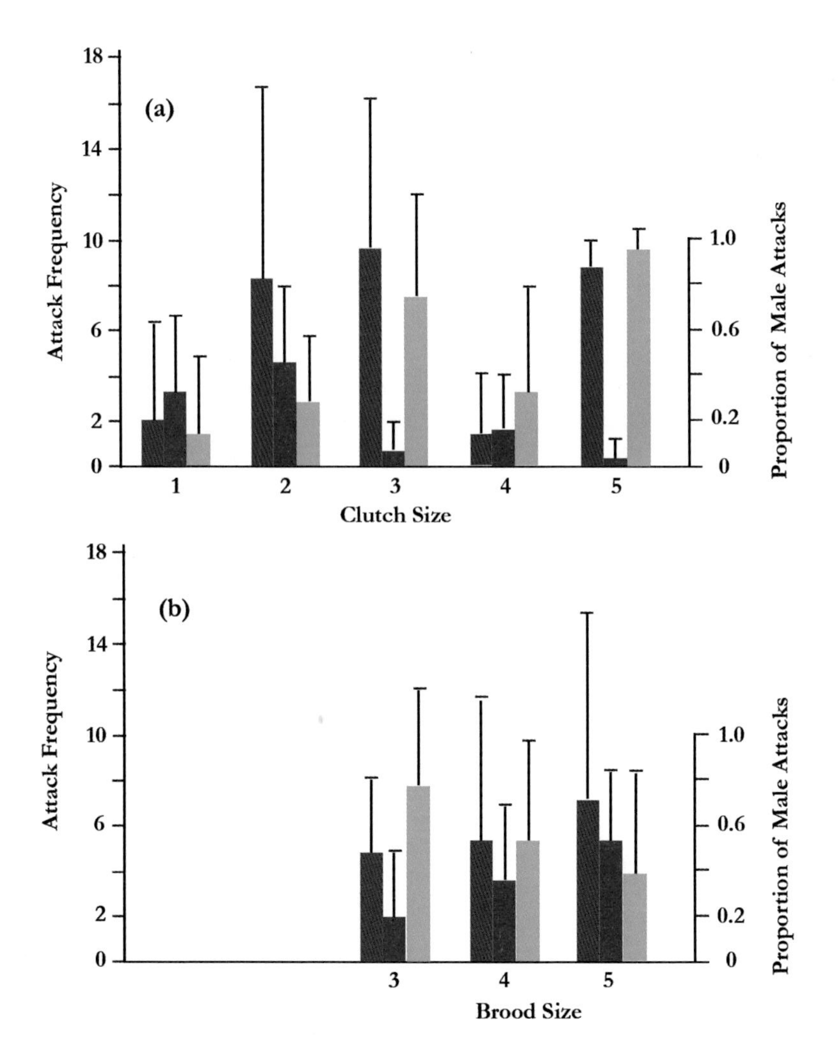

Figure 23
Attack frequency on a potential predator by breeding male and female Merlins. The right vertical axis notes the proportion of attacks which are made by the male.
a) Variation with clutch size.
b) Variation with brood size.
Redrawn from Wiklund (1990a).

the breeding attempt while there is still time to lay a replacement clutch and so attempt to salvage something from the breeding phase?

In a second part of the study (Wiklund 1990b), using the same stuffed Raven to simulate a potential predator, chicks were exchanged between nests, so that some broods were reduced to two, while others were increased to five or more, the intention being to see if nest defence was correlated to brood size. In a separate part of the study replacement clutches were observed to see if nest defence in these differed from that in first clutch nests. The results of the studies of brood manipulation indicated that female Merlins attacked more frequently when they had larger broods, the attack rate being almost triple (on average). However, some females did not react as anticipated by increasing the frequency of attacks. In the study of replacement the females attacked less frequently than females with first clutches, with the latter, on average, attacking four times as frequently.

Wiklund is cautious in assessing the results as there are other factors which might affect female behaviour (e.g. the female's condition, though measurements of the females suggested there was no direct correlation with body size, which is usually taken as an indicator of condition). However, it does seem that female nest defence is related to brood size and the likelihood of chicks surviving to reproduce. The female Merlin invests energy in reproduction (around 50-60% of body mass goes into egg production) but, it seems, it is brood size (which represents the future) rather than clutch size (which represents the past) that defines nest defence. That reasoning would also explain the reduction in nest defence for replacement nests, as it is known that the survival rate for chicks that fledge late in the breeding season is reduced.

One interesting aspect of the work of Wiklund was that it showed that predation was independent of the experience of the female, implying that the rate of predation was insensitive to nest site selection (assuming, of course, that the female Merlin is primarily responsible for choosing the nest site). The majority of Swedish Merlins breed in trees. If the likelihood of an individual nest being predated depends primarily on the chance of a predator finding it, then assuming an even distribution of predators and that all trees are equally accessible to thieves, there is little the female Merlin can do to ensure nest safety other than find the highest, but least conspicuous, stick nest in her mate's territory.

In a further study in the same study area Wiklund (1995) looked at the effect of predation on the LRS (Lifetime Reproductive Success) of female Merlins. In a study of 92 females Wiklund found that nest predation accounted for 31.1% of the variance in LRS: only breeding life span was a larger contributor. In a later study Wiklund (2001), using data covering the years 1988 to 1994, found that the annual predation from nests was 33.6% (range 16%-48%), meaning that in the worst year almost half of nests were lost to predation. However, he also found that there was no correlation between the predation losses in one year and the number of breeding pairs in the next, again implying the critical importance of habitat to breeding Merlins.

Breeding Territory and Mate Fidelity

Following fledging, juvenile Merlins move away from the natal area. In a study of the urban-dwelling Merlins of Saskatoon, Canada, Lieske *et al.* (2000) ringed nestlings and trapped adult falcons across a wide area of the city (122km^2, an expanse that was covered 'on foot, by bicycle or by vehicle'). By recapturing ringed birds, data on dispersal (and also on annual survival – see *Chapter 10*) were acquired which were then analysed after allowing for the potential statistical effect of their sampling – e.g. are all falcons equally likely to be captured/recaptured? Lieske and co-workers collected data over the period 1985 to 1996, ringing a total of 1354 Merlins (153 adults caught as breeding birds, the rest ringed as local nestlings). Of the birds captured as adults 14 males (of 43 – 32.6%) were recaptured in the following year, while 50 females (of 110 – 45.4%) were recaptured, implying a significant degree of territorial fidelity. Of the birds ringed as nestlings 10.2% of males and 4.0% of females were recaptured as adults in later years. However, the survival rates for juveniles, measured as the probability of being recaptured in the following year were 23% for males and 5.5% for females. These data could not distinguish between emigration and mortality and so the figures (extremely low in the case of females) do not reflect true survival, but, rather, survival and return to natal area, i.e. female Merlins are much less likely to return to their natal areas and, by implication disperse further after fledgling.

In an earlier study of the Saskatoon Merlins, James *et al.* (1989) had noted that on average males moved 4.1±2.9km between natal nest and breeding place, while females moved 3.0±1.4km, but the numbers involved were small, and four ringed females had moved considerably greater distances, up to 259km, suggesting that female dispersal might actually be greater than male, a view supported by the fact that more male birds ringed in the city were trapped in the city (41 of 68 – 60%) than were females (35 of 89 – 39%). Greater movements of female Merlins were also noted in breeding dispersal, i.e. the distance travelled between breeding sites in consecutive years, which were an average of 1.1±0.9km for 15 males, and 2.3±1.8km for 29 females, a difference which was statistically significant. These distances are quoted again in the work of Espie *et al.* (2004) which noted that for both male and female Merlins the move from one nest site to another invariably involved a switch to a site of better 'quality'. Quality is invariably difficult to define, the more so in this case as it is not possible to know the parameters which influence Merlin choice. This seems particularly the case in urban Saskatoon where prey and nest sites are abundant. However, when Espie and co-workers looked at the frequency with which some territories and nest sites were utilised it was clear that some were more favoured than others. They therefore used this usage factor to rank the 'quality' of sites. What then became apparent was that when both male and female Merlins moved site they moved to one of higher quality. For females this switch had little effect on either breeding success or survival to breed again. However, for males switching to higher quality sites was positively correlated to LRS. But, interestingly, males that did not breed in their first year did not gain higher quality sites in the following year than first year breeders. As Espie *et al.* (2000) showed that male Merlins that delayed breeding until their second year had a higher LRS, it is clear that those males that delay breeding are gaining experience and so trading longer term survival and LRS against an earlier chance at breeding.

The suggestion that females disperse farther than males was confirmed by Wiklund (1996) in his studies in two Swedish Arctic National Parks. Wiklund found that on average females dispersed further (4.8±6.7km, range 0–32km) than males (1.7±1.9km, range 0–9.4km) and only females travelled more than 10km (Fig. 24). Fidelity to a territory was higher in males than in females, Wiklund noting that 7 of 49 males bred in the same territory throughout their breeding life (of 2–4 seasons), while only one of 45 females did so: that female bred in the same territory through her 5 season breeding life. However, 13 other males and six females used the same territory on more than one occasion during their breeding life. One interesting finding of Wiklund's study was that males breeding in areas of high density tended not to disperse far, and that older males tended to breed in high density areas whereas younger males tended to breed in less densely populated areas. This finding seems at odds with that of Lieske *et al.* (2000) (see *Chapter 10*) who suggested that for urban-dwelling Merlins high population density caused higher male mortality. However, it is not clear whether 'high density' was comparable in the two study areas, and what the influence of the trade between survival and reproductive success was in each case.

Wiklund's finding regarding high male fidelity to a breeding territory was also confirmed by Morozov *et al.* (2013) in their study of Asian Merlins, noting that on the tundra the distance between nests in successive years averaged 2.7km, but could be as little as 100m, movements which the researchers considered made the sites permanent or semi-permanent: one area of peat bog in the study was used in 18 successive years. The Steppe Merlin was considered more conservative than its northern cousins with records of sites within the same wood being used in up to nine successive years. Karyakin and Nikolenko (2009) also record *F.c. lymani* occupying the same four sites in two valleys of their study area every year for nine successive years. As Morozov *et al.* (2013) note, such long periods of usage suggest permanent, traditional, sites, consistent with the view of Rowan (1921-22 Part I) and Newton *et al.* (1978) noted in *Chapter 3: Habitat.*

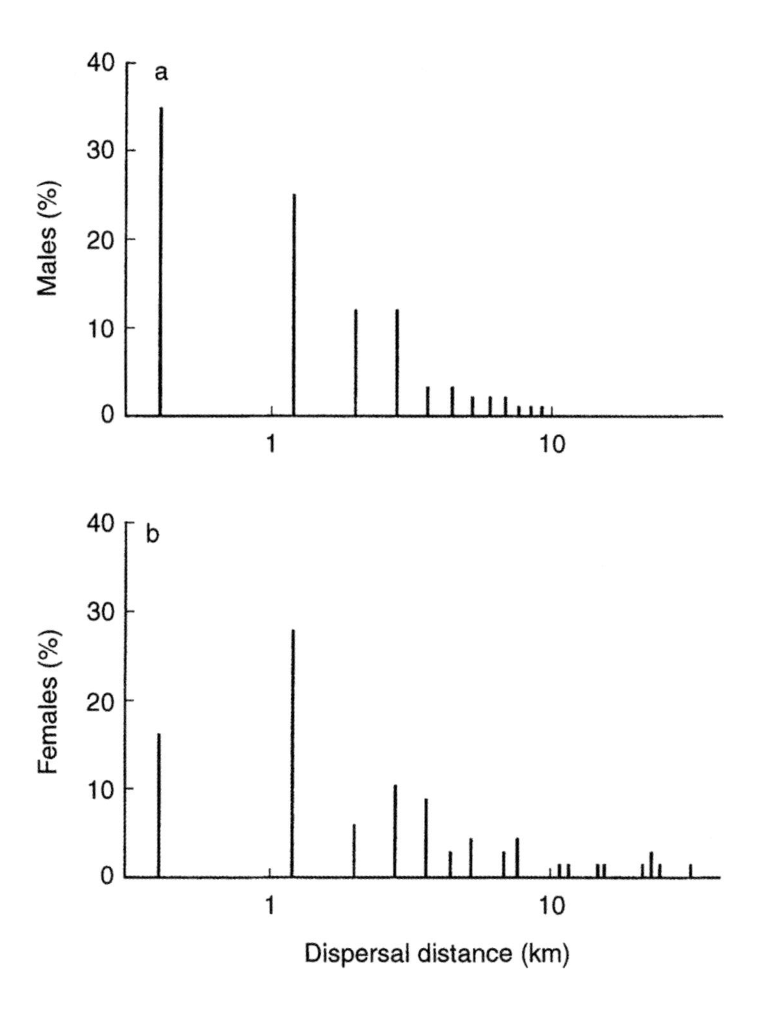

Figure 24
Dispersal distances for (a) male and (b) female Merlins between their breeding sites in successive years. The horizontal axis is log (distance) with distance measured as 0-799m, 800-1599m etc.
Reproduced from Wiklund (1996).

Site fidelity was also confirmed in a study of the urban-breeding Merlins of Saskatoon (James *et al.* (1989), Warkentin *et al.* (1991)). In the earlier study 41% of males (N=17) returned to the same nest site area, while in the later study the percentage was higher, 61% (N=28). These figures compare to 20% for females (N=49, James *et al.*) and 28% (N=39, Warkentin *et al.*). In the later study one male returned to the same nest site area for four consecutive years, but one female returned for five consecutive years. These figures might imply a high degree of mate fidelity, but Warkentin and co-workers found this was not the case, and that the likelihood of a pair that had mated in one year breeding together the following year was no better than would be expected by chance, only 20% of pairs (N=60) being the same two birds in successive years. However, males who maintained the same territory were more likely to mate with the same female than those that changed territories, a result which reinforces the random nature of pairing as it is clearly more likely that pairs will involve the same birds if nest site fidelity is reasonably high in both sexes. For that reason the likelihood of female Merlins having the same mate in successive years was relatively high for those that returned to the same nest site area (71%, N=7), but unlikely if they moved area (9%, N=11). One pair which stayed together in successive years actually moved nest site, the only result which could be construed as implying mate fidelity. The overall rate of mate switching was 68% (13 of 19 pairs in which both birds were found in consecutive years had switched mates). Warkentin and co-workers hypothesise that mate switching is a response to migration and high annual mortality, as a returning male (males normally arrive at breeding sites before females – **Box 24**) has to decide whether to mate with an available female or wait for a returning mate, when that mate might actually not have survived to return.

In their study, Warkentin *et al.* (1991) also measured the productivity of the Merlin pairs and found no significant difference between that of the same pair breeding in successive years, and no significant difference in pairs which changed mates in successive years.

On the basis of the results on *F.c. richardsonii* from Saskatoon it seems the imperative to breed makes the decision for the male, but in the nine-year study of *F.c. aesalon* in Sweden Wiklund (1996) noted nine males and ten females who retained mates during two or three consecutive years: one male bred with one female for two consecutive years, then with a different female for three consecutive years. While Wiklund's study involved a total of 220 birds, implying that 19 is a small sample, the number which bred on two occasions was lower at 94, and the number that bred in three years lower still at 46, so the same mate data does imply a degree of mate fidelity in contrast to the work of Warkentin *et al.* (1991). Interestingly, Hodson (1976) studying *F.c. richardsonii* in rural Alberta, Canada, had results which were closer to the rural population of Wiklund (1996), with 9 of 12 males (75%) recaptured in the second year of study being on the same territory, but only 2 of 20 females (20%).

A further complication is added by the study (on *F.c. aesalon*) of Wright (2005) in Yorkshire, north-east England. Wright confirmed Merlin territory/nesting area fidelity, but noted that this did not necessarily mean nest site fidelity, pairs tending to nest within a given area, but sometimes

Box 24
While it appears logical that the reason for males arriving earlier than females is in order to claim better territories and therefore to be potentially more attractive to females and so enhance their reproductive chances. However, this hypothesis is still debated, the main point of contention is whether the cost to males arriving early (as this may require migration and arrival when resources are not optimal and so can threaten survival) – termed viability selection – is compensated by increased mating opportunities (sexual selection). For a good review of how the factors compete see Morbey et al. (2012).

choosing different nest sites within the area. This finding is consistent with the study of Newton *et al.* (1978) who found that the fidelity to an individual nest site, as opposed to a general nest area, was not statistically significant. However, Newton and co-workers found that in some areas Merlin pairs nested on the ground where trees with stick nests were available, implying that in most cases fidelity to a nest area overcomes the apparent (and debated) extra security of utilising stick nests which were not within the habitual general nest area. But it must be emphasised here that the situation, in terms of nest site fidelity as opposed to nest area fidelity, is different for ground nesting and tree nesting Merlins. While the latter require a stick nest and these are usable only over finite times, ground nest sites can be used repeatedly. As already noted (see *Chapter 3: Habitat*) Rowan (1921-22 Part I) notes continuous occupation of a specific site over a period of at least 20 years, while Newton *et al.* (1978) present evidence of sites that have been in used for almost 100 years. However, while these accounts imply long-term species fidelity to territories/nesting areas, and it may be inferred from them that pairs of birds which have successfully bred in a particular area may return the following year, there is limited direct evidence to support mate fidelity.

Wright (2005) captured 6 male and 10 female Merlins that had rings which enabled their nest of origin to be established. Only one of the males, and none of the females had been hatched close to where they were recaptured: on average the females had travelled furthest from their natal area (58.1km, range 18.9-105.0km) (**Box 25** overleaf), males travelling an average of 30.6km, range 2.7-63.3km). Wright also recorded the number of times birds were recorded returning to their breeding sites. Over a 10 year period (1993-2002) only two males (2 of 11 – 18%) were discovered to have returned in successive years. In each case they occupied the same territory in those years, though the nest sites used moved (by 360m and 700m). Female Merlins seemed more likely to return to areas where they had successfully bred. Four of 33 females were recorded in successive years, though in each case they had moved to adjacent territories. Two of the 33 were recorded three times, one using the same territory in successive years, then moving to a territory 4km away: the other female used adjacent territories on three occasions in four years, being absent from the area for one year. Two of the 33 females used the same territory in four successive years, while a third female used two territories, breeding in each for two successive years. Finally, one female was recorded in two successive years, was then absent for a year, then returned, could not be assessed in the next year when access to the moor was stopped due an outbreak of foot-and-mouth disease, but was recorded again in the sixth year. Over the years of recording this female used three adjacent territories.

Box 25

The British data of Wright (2005) is consistent with that of Marsden (2002) from north-east Scotland who noted, using DNA profiling, that of 185 females 82% bred in a particular nest area once only: 10% of females nested twice, with 5% found three times, 2% four times and 1% five times. Data for 96 males found 74% breeding in the same nest area once only, with 19% twice, 5% three times, 1% four times and 1% five times. Marsden also found that if a female returned to a nest area she would mate with the same male on 39% of occasions, with a different male on 11% of occasions: on the remaining occasions the same male would mate with a different female (22%) or both male and female would have changed (29%). British data is also consistent with that from Iceland in Nielsen's (1986) study of *F.c. subaesalon*. Nielsen trapped live birds as well as recovering dead birds which he had previously ringed as nestlings. Of the live birds, five females had moved 22km, 22km, 41km, 62km and 66km from their natal sites: no live males were re-trapped. Of the birds recovered dead, two females were 38km and 78km from their natal sites, one male had moved 5km, and two birds of unknown sex were at distances of 22km and 23km. Nielsen also trapped birds to see what the fidelity was to nesting territories. Of 16 female Merlins, eight (50%) were at the same territory as the previous year, while four had moved: of eight males, four (50%) were on the same territory as the previous year, while two had moved. The remaining birds were not re-trapped. The falcons which had moved territories had moved less than 10km. One male-female pair which had bred successfully in one year, mated again the following year, but on a territory which was 2km away: the pair's original territory was occupied by a new pair. One other pair which had bred successfully had occupied the same territory again.

Further work will be required to decide if the observed differences between the behaviour of rural and urban Merlins, and the differences observed in certain rural populations, are real or derive from small sampling numbers.

Usually the female Merlin feeds the nestlings, but at this nest of Eurasian Merlins in Scotland the male did his share of domestic duties.

Movements and Winter Grounds

The majority of Merlin populations are migratory, moving to and from their breeding grounds in order to avoid the rigours of the northern winter. Below, each sub-species is dealt with in turn, before more general aspects of migration, resident populations and the behaviour of the wintering birds are considered.

F.c. columbarius

Nominate birds are the most migratory of the various sub-species, some birds actually reaching the Equator in northern South America. Birds from Alaska and western Canada migrate south along the US coast to Baja California and the nearby west coast of Mexico (e.g. Enderson *et al.* 1991), and also to the coasts of the Yucatan Peninsula), to Central America and South America, wintering all along that migratory route, as far south as Ecuador and northern Peru. However, some Alaskan birds only travel as far as Kodiak Island. In South America wintering Merlins are uncommon everywhere, though they are relatively common on the Dutch Antilles where they feed on shorebirds. There are also three definite records of a nominate bird finding its way to Europe. A juvenile male was found dead in Iceland (Pétursson *et al.* 1992), a female/juvenile was photographed on an island off County Clare, Ireland (Garner 2002) and another juvenile was seen in the Azores (Crochet 2008). There is also the specimen (Knox 1993), an adult male, shot by Meinertzhagen on South Uist, an island of the Outer Hebrides off Scotland's western coast, in November 1920 which has been generally accepted as a nominate bird.

Schmutz *et al.* (1991) present data on the recovery of Merlins ringed in western Canada. This study's findings established that the autumn migration (from Alberta and Saskatchewan) begins in September/October, though the recoveries suggest that some birds are resident or move only relatively short distances (30 of 44 recoveries were from Canada). Six of 11 recoveries in the USA were in Colorado, suggesting that this might be an important wintering area for western North American Merlins. However, one bird ringed in the Yukon close to the Arctic Circle was recovered from Costa Rica. Schmutz and co-workers suggest that this bird, and another recovered in northern Mexico are indicative of 'leap-frog' migration (**Box 26**) as the two had flown over both breeding and wintering ground of *F.c. richardsonii.*

In eastern Canada there are, as noted above, small resident populations in several areas. Those Merlins that are migrant head south along the

Box 26
Leap-frog migration is the term for northern migrating birds which travel faster and further than southern migrators, the entire journey of the latter being encompassed within the journey of the former.

eastern seaboard of the United States to Florida, where they are most commonly found on the Keys, with some crossing to the Bahamas (e.g. Buden 1992a, 1992b) and northern Caribbean islands (Cuba, Haiti/Dominican Republic and Puerto Rico). Large concentrations of Merlins are seen at points along the mid-Atlantic coast: at one point 867 Merlins were seen on a single day (Wheeler 2003a). Wheeler also notes that daily flights may be offshore and so cover large distances over water, the birds coming ashore to feed and roost. Some birds, particularly females, may make short migrations only if they find areas where prey is abundant. Adults return earlier than first year birds, heading north from late February through to early April: Merlins arrive in Alaska from late March to mid-April (Laing 1985).

Male Taiga Merlin, photographed in October, New Jersey.
Julian Hough.

F.c. richardsonii

The sub-species represents the largest resident population in North America, many birds wintering within their breeding range in urban areas where prey remains plentiful. Migrant birds travel south through the central United States to reach Mexico and central America. Wheeler (2003b) states that Merlins did not winter in Canada until the House Sparrow became established in 1900 and was even then rare until the 1950s when Bohemian Waxwings began to appear in urban areas in large numbers as a consequence of the availability of food provided by maturing berry bushes and trees planted by human gardeners. Merlins are now resident or partially migratory in many urban areas and also winter on the central and southern Great Plains, particularly Colorado, a northward expansion and new habitat occupation which, it has been suggested (James *et al.* 1987a), has resulted from an increase in population, coupled with an increase in prey availability as a consequence of the increase in Bohemian Waxwing numbers. However, while urban populations are resident or partially migratory there are also fully migratory populations. The migrations of these start in early September. Adult males migrate northwards first, leaving their wintering grounds in February/early March.

On passage and in winter in the United States Merlins are most often seen hunting on arable or other farmland, but sightings are uncommon as in general the falcons avoid settlements, and even farms.

F.c. suckleyi

There is less data on the movements of this sub-species than on the other, more numerous and more wide ranging, North American sub-species, but it seems that most birds are resident, though some are migratory, wintering along the US coast south of Seattle as far as southern California, New Mexico and northern Baja California (Wheeler 2003b).

Haak (2012) notes that all three Nearctic Merlin sub-species are present during the winter in south-west Idaho.

F.c. subaesalon

The Icelandic bird population comprises both resident and migratory birds, though the resident birds are few in number – Nielsen (1986) saw only one during his study period, but notes that they have been seen infrequently in all areas of the island. Formerly, when it was assumed that the sub-species *F.c. subaesalon* and *F.c. aesalon* could be distinguished by wing length (see *Chapter 3: Distribution*) birds found on Fair Isle (which lies between the Orkney and Shetland islands, to the north of the British mainland) were believed to have originated in Iceland, but it is now considered that these birds were migrating from the Shetlands. Assumed *F.c. subaesalon* birds were also recovered on mainland Europe, but these are now also considered to have arrived from Britain. However, ringed Icelandic birds have been recovered, in very small numbers, in Norway, France, Belgium, Spain and the Azores (in the case of the latter the falcon's leg ring was returned in an envelope posted from the Azores, but there was no information on exactly where the bird was recovered or the circumstances of the recovery). There is also the curious tale of the Merlin in the Zoological Museum of Amsterdam University which was presented to the museum after being recovered on a ship off the Bahia coast of Brazil. This bird has been ascribed (van Bars-Klinkenberg and Wattel 1964) to the Icelandic form *F.c. aesalon*. Given the difficulty of being absolute in terms of differentiating some sub-species it seems much more likely that this recovery is evidence of a very long migration flight by a Nearctic Merlin rather than an extreme flight by an Icelandic bird.

Other Icelandic birds have been recovered from the Faeroes, where the population is partially migratory. One bird was also apparently recovered from Greenland, though attempts to confirm the circumstances of this

Female Icelandic Merlin. Migrating Icelandic falcons have to cross a large expanse of unpredictable ocean.
Sindri Skúlason.

have failed, with Icelandic researchers dubious as to its veracity. The majority of recovered Icelandic birds have been found in western Britain and Ireland. One bird from Iceland, captured and ringed in Britain was discovered dead aboard a ship close to Iceland, presumably having died during an attempted return migration to its breeding grounds.

Nielsen (1986) noted that some Merlins were migrating from Iceland in August or early September (one bird recovered in Ireland on 20 September was 1600km south of the natal site where it had been ringed), but others were later with migration continuing through October: the latest recovered bird assumed to be migrating was found on 24 November 295km from its natal site. Icelandic migrants tend to fly via Vestmanaeyjar (the Westmann Islands) off Iceland's southern coast and from there to the Faeroes. In spring, birds start arriving on the southern Iceland coast in early April, and in the northern breeding areas a week later.

The Faeroese population is partially migratory, though recoveries of birds ringed on the Faeroes are few: most of these have been recovered on the islands themselves, though one bird was recovered in County Sligo, Eire. Resident Faeroese birds are assumed to be joined by Icelandic Merlins and, perhaps, by birds from Scandinavia. One Merlin ringed in Scotland in June 1996 was recovered in June 1997 on a ship to the east of the Faeroes. Observations suggest that most resident birds are first-years as adults are rarely seen in winter. The number of Merlins tends to be highest in September, a period which corresponds to the arrival of migrating Northern Wheatears from Iceland and Greenland. At that time it is possible to see ten or more juvenile Merlins hunting Wheatears at the southern tip of the archipelago (pers. comm. J-K Jensen and S Olafson).

F.c. aesalon

Most ringed British birds have been recovered within Britain, suggesting that British Merlins which are not resident move relatively short distances (an average of about 100km), chiefly to lower elevations and coastal areas, particularly estuaries, where personal observation suggests their presence usually causes panic and chaos among populations of waterfowl as well as shorebirds (the most likely prey). However, British birds have been recovered from continental Europe. The recovery of ringed birds has also shown shipboard curiosities: on ships near Norway, north of the Faeroes and south of Ireland. Wright (2005) records the recovery of only three adult birds ringed at his breeding sites during the period 1987-2002. One male travelled 264km, while two females travelled 196km and 287km. In addition, 21 of 267 ringed chicks were recovered. The average distance travelled between fledging and recovery was 119.2km (range 2.5-224.0km) for males and 76.5km (6.5-119.0km) for females, though the latter excluded a single female which travelled 1239km (from northern England to Navarra in northern Spain) in 91 days. Movements in Britain begin in September, with birds returning to their breeding areas in April.

In general, the recovered birds in Wright's study had travelled south, though some had travelled east or west to reach the coast. This is consistent with the finding of Newton *et al.* (1978) in a study in which 394 hatched Merlin were ringed as nestlings. Of these 19 were recovered before the onset of winter, and of those 18 had moved less than 100km from their nest site, but in random directions, initial dispersion seemingly a search for an available hunting range rather than any early migration.

The female bird recorded by Wright (2005) which travelled over 1000km is not unique. Other long-distance travellers (Heavisides 1987) include an adult male ringed on Fair Isle, off Britain's northern coast, which was trapped alive in France having travelled 1767km in just 23 days, and a bird ringed as a nestling in Cumbria, north-west England on 2 July 1938 and shot in France on 20 October 1938 after travelling 1178km in around 100 days (the age of the bird at ringing is not given so it is not clear when it fledged).

Merlins which vacate their summer territories are mostly found on estuaries or other coastal regions, with a spread of observations that covers most of the British coastline.

Most Fennoscandian birds migrate, but some individuals do not, winter populations being known from Norway and Estonia, and small numbers being assumed to stay in southern Sweden and the other Baltic states. Merlins ringed in Fennoscandia have been recovered from mainland Europe as far west as Spain, and these may be responsible for those few falcons known to winter in western North Africa having crossed the Mediterranean via Italy and Malta or the Straits of Gibraltar. (See Thiollay (1977) and Corso (2001) for details of the small numbers of Merlins crossing at Gibraltar and Sicily respectively during spring migrations.) Very few birds ringed in Fennoscandia have been recovered in Britain, those birds seen seeking rest or shelter on North Sea oil and gas rigs now being believed to be British in origin (Heavisides 2002 and reference therein). This lack of Fennoscandian birds in Britain is consistent with observations of migrating raptors at Falsterbo in southern Sweden, from where the sea crossing to Denmark is just 15kms. The autumn migration may begin as early as August, but continues into September. For falcons wintering in north Africa the return journey may start as early as February with Arctic breeding birds not arriving until May.

Merlins from the Baltic states, Belorussia and western Russia may also migrate through northern and central Europe, but may head south to join migrants from further east in Russia to reach Greece and Turkey, and the states on the western and northern side of the Black Sea. Morozov *et al.* (2013) note that Merlins ringed on the Kola Peninsula have mostly been recovered from Germany, Italy (including Sicily) and France. Morozov and co-workers also suggest that while the ringing data is limited, it does imply leap-frog migration (see **Box 26**), northern Merlins seemingly wintering much further south than their southern ringed cousins. However, it seems that some northern birds do not migrate, wintering Merlins occasionally being found in St Petersburg where they were hunting passerines in parks, public gardens and cemeteries: Merlins have also regularly been seen during winter in Moscow in recent years and they have also overwintered near Rostov in the winter of 2005/2006 when there was a winter irruption of White-Winged Larks on which they hunted (Morozov *et al.* 2013).

F.c. aesalon also winter in Kazakhstan, Turkmenistan, Uzbekistan, and further south in the Arabian Peninsula. They have also been seen during winter in Mongolia, northern Pakistan and India. Departure and return times are similar to those *aesalons* breeding in Scandinavia.

F.c. pallidus

Steppe Merlins migrate to countries around the Black Sea, to Arabia, and to the countries from the Caspian Sea to the Tien Shan Mountains. The falcons also reach the Tibetan plateau and from it northern Pakistan and India.

Meinertzhagen (1924) records having seen three sub-species of Merlin – *F.c. aesalon*, *F.c. insignis* and *F.c. pallidus* – forming small parties during one winter in northern Iraq, and that the birds also roosted together. Meinertzhagen does not mention numbers, but says that he 'obtained five specimens' these being one *aesalon* female, one *insignis* female, and three *pallidus* (one male and two females) though the data on wing lengths suggests four *pallidus*.

In general *pallidus* begin the autumn migration in September, much as for the other sub-species. However, they tend to arrive early in their breeding grounds, some as early as mid-March, though most arrive in April. These later birds may be those which have wintered in the northern Indian sub-continent as these do not leave until March, with some stragglers not moving until early April. Further details on observations of migrating Steppe Merlins can be found in Morozov *et al.* (2013).

F.c. lymani

The population of the Central Asian Merlin is believed to be essentially sedentary, though some movement may occur as a result of prey migrations.

F.c. insignis

The wintering area of the East Siberian Merlin is almost as vast as that of

F.c. aesalon, with falcons having been seen as far west as Jordan, Iran and Armenia, while to the east they have been observed in China (the largest fraction of the population overwinters in southern China), South Korea and Japan. Between these extremes the falcons overwinter in southern Russia and Mongolia, and reach the eastern Tibetan plateau from where they cross the Himalayas to reach northern Pakistan and India. See also the comment of Meinertzhagen (1924) in *F.c. pallidus* above.

Migration timing has not been well studied, but it seems that northern falcons leave their breeding grounds as early as 12 August, though some are known to have been still on the Taimyr Peninsula in early September (Morozov *et al.* 2013). Spring migration starts in mid-March and continues through April with falcons reaching Arctic breeding grounds in May. Further details on the times of observation of migrating falcons is given in Morozov *et al.* (2013).

F.c. pacificus

The Pacific Merlin winters in Russia's Primorye region (close to the Chinese border on the Pacific coast), in China as far south on the mainland as the latitude of Taiwan's southern tip, as well as in Korea and on the island of Taiwan, and in Japan, from southern Hokkaido southward (Brazil 2009). Primorye has a breeding population of falcons, so these may well be resident, migrants swelling the winter population. There is little information on the timing of migration fights, but such information as is available (Morozov *et al.* 2013) suggests that the spring migration is over an extended period, Merlins being seen crossing Sakhalin Island during the whole of May. The autumn migration appears late in the northern part of the range, falcons not leaving Anadyr until mid-September, these timings again suggesting leap-frog migration. In Primorye, these late departing birds are not seen until late October or even into November.

Migration Flights

Kerlinger *et al.* (1983) studied data on the autumn migration of raptors observed from cruise ships in the waters off north-eastern USA, from Nova Scotia south to Cape Hatteras, North Carolina. While in general Ospreys were seen further from land than other raptors, there were sightings of Peregrine Falcons and Merlins (sub-species not specified, but probably *F.c. columbarius* on the basis of the location of sightings) often at distances which meant that the falcons were not able to see either the coastline to the west or the coast in the direction in which they were heading (95.7% of observed Merlins were out of sight of land). Rather than following the coast as most of the observed Merlins did, these birds seemed to heading across the ocean from Nova Scotia towards Cape

Female Pacific Merlin. The falcon was photographed on the shore of the Sea of Okhotsk during migration. It is feeding on a Yellow Wagtail.
Igor Dorogoy.

Hatteras, a distance of more than 1000km. While it is possible the birds were breaking their journeys by heading inland, one bird was seen 300km off Cape Cod (though the mean distance from land was 87±56km, N=25). And one was seen to be consuming a small bird on the wing, presumably taking opportunistic advantage of an exhausted migrating passerine.

For further details of the spring migration of Merlins along the coast of New Jersey (Cape May Point and Sandy Hook) see Clark (1985) who gives details of timings (e.g. that females migrate earlier than males in autumn, later than males in spring), age and sex ratios, and the weights of adult and juvenile birds when captured (the falcons were heavier in spring than in autumn).

Seeland *et al.* (2012) monitored autumnal migration patterns at Hawk Ridge, Duluth at the south-western corner of Lake Superior. They found that there were marginally different factors influencing the migration flights of high soaring raptors and lower flying hawks and falcons, the latter (including the Merlin) being influenced by not only wind direction, but antecedent wind direction (i.e. the direction over preceding days), and also time of day, ambient temperature and seasonal interval (i.e. whether

Other migration/winter Merlin prey. Least Sandpiper (*above left*), Dunlin (*above right)* and Redshank (*below*).

the birds were early or late in the migration season). Seeland and co-workers consider their study provides valuable information at a time when wind power development is gathering pace since it is known that birds are at risk from the blades of wind turbines, so that an understanding of the interaction between migrating birds and the landscape is vital.

The study of migrating raptors at Falsterbo, southern Sweden, by Kjellén (1992) confirms that Merlins (in Kjellén's case *F.c. aesalon*) were less daunted by crossing open water than some other species, the falcons not concentrating in large numbers at his study area, instead migrating on a broad front. However, it should be noted here that, as we have seen, very few *F.c. aesalon* are willing to cross the North Sea, so Kjellén's view is more likely to be applicable to the relatively short crossing from southern Norway and Sweden into Denmark. Merlins do cross at Falsterbo, however, with some idea of the different choices made by the raptors being seen in the numbers counted: while Kjellén counted 1105 Merlins in the five autumns of his study (1986-1990) he counted over 72,000 Eurasian Sparrowhawks, and over 56,000 Common Buzzard. Even allowing for the differences in total Fennoscandian population of the three species, the difference in migration route choices is clear.

Kjellén studied the composition of the populations of the migrating Merlins, and noted that adult females migrated first, juvenile birds of both sexes migrating later and adult males migrating even later (Fig. 25). The difference in median dates of migration for the three groups were statistically significant. This migration behaviour is similar to that of the Eurasian Kestrel, but differs from that of the Hobby and Peregrine where adult birds migrated at the same time, but ahead of juveniles. As noted above, similar differential migration between adult male and females has also been seen in Merlins in eastern North American (Clark 1985 at New Jersey) and in juveniles in western North America (Hull *et al.* (2012) in California, though this provides confirmation only for juvenile Merlins as the sample size for adult falcons was too small for meaningful statistics), while a study by Wittenberg and Smith (2009) showed that there was no differential migration in autumn in juvenile Merlins sampled in the Florida Keys. Wittenburg and Smith collected feathers from trapped birds and used the ratio of deuterium (a stable isotope of hydrogen) to hydrogen in the feathers to define the origin of the birds as it is known that the ratio

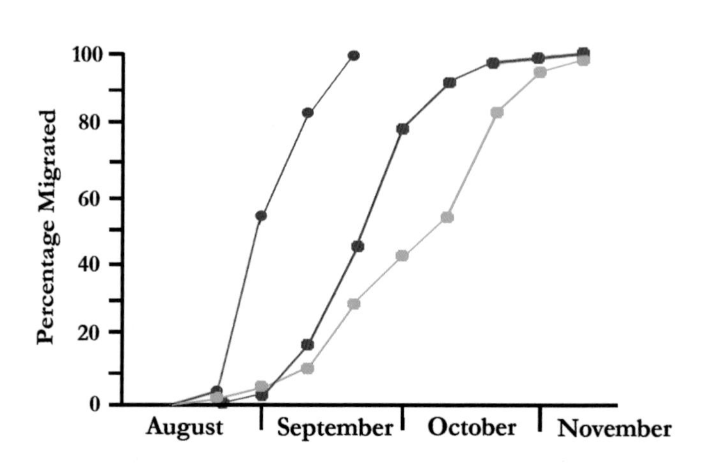

Figure 25
Autumn migration of Merlins at Falsterbro, Sweden. The data were based on numbers collected in 10-day periods. Percentage migrated is the cumulative percentage of birds which had migrated at a certain date.
Redrawn from Kjellén (1992).

varies across North America. The results confirmed the absence of juvenile differential migration. However, a further study using the same technique has thrown this result into doubt as it suggested a systematic error in the natal origin of the Merlins (Wittenburg *et al.* 2013), with all sampled juveniles having apparently been hatched further south than the known breeding range of the species.

The reasons for differential migration are still discussed. Kjellén (1992) considers the most likely explanation involves the timing of the moult, with females having completed the moult during incubation and so are ready to migrate earlier than the males who must wait until later to moult as they are the primary supplier of food to the nestling. The moult hypothesis does not contradict the migration pattern seen in those falcon species in which the males and females depart at the same time, as these birds are long distance migrators and arrest the moult in autumn prior to migration, completing it at the winter quarters.

However, Mueller *et al.* (2000) have questioned Kjellén's data as regards the differential migration of Merlins. Mueller and co-workers trapped 23,000 raptors of 10 species between 1953 and 1996 at Cedar Grove, Wisconsin, the trapping allowing the aging and sexing of the birds to be reliably accomplished (one criticism of Kjellén's data by Mueller *et al.* (2000) relates to the difficulty of ageing Merlins in flight). Mueller and co-workers found that adult males older than two years migrated later than older adult females, but that there was no significant difference in the timing in the sexes for two-year old adults. Adults migrated later than hatch year birds, and within hatch year birds males migrated later than females. The criticism of Kjellén by Mueller and co-workers rests on the ability of the former to distinguish females and juvenile birds adequately in flight which throws into doubt Kjellén's contention that females precede juveniles, though it does not contradict Kjellén's view of the validity of the moult hypothesis, as the sample size of Mueller and co-workers for two-year old adults, particularly females, was small. As noted above, the study of Hull *et al.* (2012) in central coastal California, also confirmed that first-year females migrated before first-year males.

Non-migratory Populations

In one of the series of studies on the urban-dwelling Merlins of Saskatoon, Warkentin *et al.* (1990) noted that in addition to its resident population (which averaged 30.5 birds during the four winters of the study) the city had a large wintering population. The aim of the wintering study was to shed light on the four hypotheses that have been put forward to explain why such partial migration might develop. Briefly the hypotheses are: *body size*, which proposes that smaller birds migrate as those with larger bodies can endure the lack of food that residing in the breeding area demands; *dominance*, which suggests that the population hierarchy means that dominant birds can out-compete subordinate birds, forcing them to move to avoid starvation; *arrival-time*, which maintains that breeding birds obtain better territories by arriving first in breeding areas, and the most efficient way of doing so is not to leave; and *genetic*, which argues that the migrate/stay decision is genetic.

In their study Warkentin and co-workers found no significant body size differences in the population of wintering Merlins. Furthermore, the age structure and sex ratio of the trapped Merlins did not support the dominance hypothesis, while the age structure did not support the arrival-time hypothesis. However, it was found that 19 of 20 birds trapped in winter had at least one parent that had also been captured in the city in winter, while only 4 of 13 birds which were hatched in the city, returned to breed there the following year and were trapped during the winter had at least one overwintering parent. The research team also consider that this fraction (31%) could have been an overestimate because of detection rates. While further work is obviously necessary before final conclusions can be reached, the results of this elegant study would appear to confirm that the decision of a bird to migrate or remain is genetic in origin.

In a further research programme on wintering Merlins in Saskatoon, Warkentin and Oliphant (1990) studied the falcons during a five-year period in the 1980s. Merlins were trapped and marked with colour-coded leg streamers, with a number of birds being fitted with radio transmitters. The home ranges of the Merlins could be established from the radio-tracked birds and showed a significant difference between adults (mean range 5.0km^2, range 2.5–8.0km^2, N=5) and juveniles in their first winter (mean range 8.5km^2, range 2.9–18.3km^2, N=3). For the adult falcons the winter ranges were separated from their breeding sites, though dispersal distances were small (a mean of 1.1km for males, 2.0km for females). These dispersal distances were close to the distances later measured to the breeding sites used the following spring, and in each case females moved about twice as far as males from and to breeding ranges. For birds experiencing their first winter, the tendency was to move further from their natal site to their winter quarters (males 2.9km, females 2.8km).

Some of the home ranges identified by Warkentin and Oliphant included sections of country outside the city limits where the Merlins hunted in farmyards and on adjacent farmland, though the bulk of the birds' time was spent in residential areas: all roosts were within the city, with adult males showing a greater fidelity to roosting places than either females or first winter birds. Warkentin and Oliphant consider the reason for the falcons moving from the city to rural surroundings, which at first seems strange given the extra energy required in flying out and back and the abundant city prey, was explained by the higher success rate in the farms. In the city the Merlin success rate was 12.9%, while in the farms it rose to 31% which Warkentin and Oliphant suggest was due to the lack of familiarity of rural prey, and so its greater vulnerability, to Merlin attack.

As previously mentioned, see *Chapter 6* and Figs. 13 and 14, the Merlins normally hunted in the morning and late afternoon. But what is interesting is the difference between the number of flights made by adult and first winter Merlins in the time after a meal (Fig. 26) which indicated that first winter birds were more active than their adult counterparts in the first hour after a meal, and very much more active as the time from the last meal lengthened. The initial activity implies an eagerness to explore their habitat (and to do so when well-fed), while the latter implies a nervousness about the likelihood of the next capture, itself an indication of lack of hunting experience.

Finally Warkentin and Oliphant (1990) looked at the interaction between Merlins who were sharing the city. This ranged from completely ignoring a second falcon to displays of aggression. The lack of response contrasts to other winter studies where the response was invariably aggressive. The difference is probably explained by the abundance of available prey, aggression being the natural response if prey is scarce, or if a second falcon ruins the hunting opportunity of the first. The aggression seen by Warkentin and Oliphant might then fall into this second category.

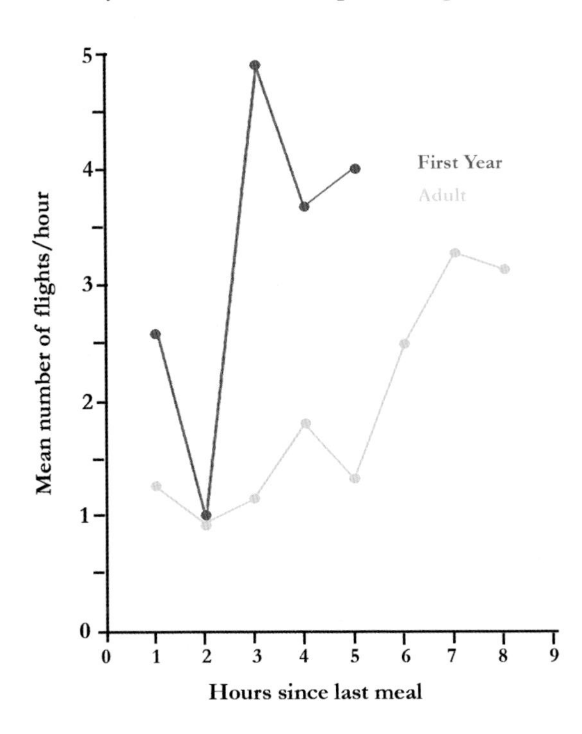

Figure 26
Mean number of flights per hour since last meal by adult and first year Merlins in winter. Redrawn from Warkentin and Oliphant (1990).

Winter Behaviour

There is little information regarding Merlin density in wintering quarters. They are generally solitary outside the breeding season, though Cade (1982) notes that juvenile Merlins occasionally migrate in loose groups which also include Sharp-shinned Hawks (*Accipiter striatus*).

In Saskatoon, Warkentin and Oliphant (1990) measured the home ranges of wintering adult and first winter birds – mean adult range 5.0km^2 (range 2.5–8.0km^2, N=5), mean first winter range 8.5km^2 (range 2.9–18.3km^2, N=3) which corresponds to a winter population density of about 16 birds/100km^2, consistent with another study in the same city by Warkentin *et al.* (1990) which measured 18.0-31.1 birds/100 km^2 (based on resident populations of 22-38 birds occupying a city covering 12,200ha). Interestingly, these wintering densities are similar to those measured in winter censuses of raptors in rangeland of El Paso County, Colorado (Bauer 1982). In that study, eight raptor species were counted as observers drove the 103km, 1.6km wide strip of rangeland which chiefly comprised the two perennial grasses Blue Gramma (*Bouteloua gracilis*) and Sandhill Bluestem (*Andropogon hallii*). Given the width of the rangeland strip it is reasonable to assume that the Merlins seen were inhabiting the entire area of 165km^2 which gives a mean density of 19 falcons/100km^2 (range 5–24 birds/km^2) across the study periods of 1962-1980.

The study of Warkentin and Oliphant (1990) on wintering, urban-dwelling Merlins also noted that there was a degree of overlap between the hunting ranges of individual birds, as was the case for the hunting ranges of breeding falcons (Sodhi and Oliphant 1992).

The wintering density of Merlins has not been measured in Britain, though it is usually assumed that the wintering population is higher than the breeding population as the influx of Icelandic birds outweighs the migration of British birds. Cramp and Simmons (1980) state that winter ranges are 'restricted' leading to marked aggression towards other species. However, Dickson (1989) states that in his experience this is not generally the case. Dickson saw only one Merlin with a well-defined hunting range during a study period of 1965 to 1984. In this case a female Merlin occupied a hunting range of about 0.3km^2 comprising rough pasture and a stubble field which was home to a flock of more than 300 Skylarks and in this one instance the restricted range did appear to result in aggression, Dickson reporting four attacks on Hen (Northern) Harriers, three on Peregrines and once on a Eurasian Sparrowhawk. Dickson speculates that the reason the female selected a specific hunting range might have been due, in part, to the winter being exceptionally cold, the permanent Skylark flock representing readily available prey, and the fact that the range lay only about 2km from a communal Merlin roost (see below) which the female apparently shared.

In a study of wintering Merlins in Northumberland, north-east England, Kerr (1989) noted that the assumption that most Merlins move from the, relatively elevated, moorland breeding grounds to coastal areas seemed valid, though with the proviso that there are many more observers in coastal areas than on the moorland in winter which has the potential to influence the result of falcon counts. Over the course of two winters (1986/87 and 1987/88) 295 Merlins were observed, 211 at, or within 3km of, the coast, with only 84 inland. Of the sightings, only 26 coastal birds were positively identified as male, while 32 inland birds were definitely male. This accords with the idea that male Merlins prefer to stay close to breeding territories in winter if that is a possibility so as to be in a good position to obtain mates and to breed the following spring. Of the 32 inland males, 18 were identified as being within a short distance of known breeding sites. Kerr notes that as the average breeding population of Merlins in Northumberland in the decade prior to the winter study was 18 pairs, this observation was further proof of the reluctance of male Merlins to move any further from breeding sites than was absolutely necessary to find enough prey to survive the winter: although on the high moorlands of the area prey is very scarce in winter, Chaffinches, House Sparrows and Starlings were available in the valleys which run down from the moors and the Merlins were taking advantage of these species.

Scott (1968) also notes an association between Merlins and Starlings, though in this case the Merlins appeared to be sharing a Starling willow copse roost (at which several other species, including pigeons and Magpies were also present), and although Scott noticed the consternation among the Starlings when the Merlins (there were up to four falcons) arrived he saw neither an attack by the falcons nor any evidence of a Merlin eating a roosting companion. However, once the Starlings left, the Merlins also left. This is curious behaviour, perhaps best explained by the Merlins predating the Starlings at certain times of day which happened to coincide with Scott's absence, though that would suggest the Starlings chose to stay at the roost despite the risks. British Merlins are also known to roost on the ground amongst dense vegetation or on low branches in woodland thickets during the winter. Merlin pairs have been seen to winter together, and communal roosting has also been reported in Britain. In Scotland MacIntyre (1936) records a communal tree roost which had been used continuously for at least 40 years in which up to 8 Merlins had been seen, and Dickson (1973) reports another observed in three consecutive winters. The roost was in an extensive area of willow trees which grew in waterlogged ground close to the sea coast, and up to five birds were seen in residence. The Merlins arrived around sunset (birds arriving both before and up to half an hour after) and would usually be gone before sunrise, up to 50 minutes before. Both 'blue' (male) and 'brown' (female or juvenile) birds used the roost and occasionally, either in the evening or the morning, the birds would engage in spectacular aerial chases. Kelly and Thorpe (1993) also noted a communal roost of Peregrines on the Isle of Man, off England's west central coast which was usually shared by Common Kestrels, Eurasian Sparrowhawks and Common Ravens, and occasionally by one or two Merlins.

In a study of urban-wintering Merlins, which had been radio-tagged, Warkentin (1986) found that the birds left their roosts before sunrise (96% of 48 birds) and returned after sunset (93% of 46 birds), but that the exact times were positively correlated with ambient temperature. Overall, it seemed that the decision to start hunting in the morning was more dependent on ambient temperature than on light level, but the decision to return to roost in the evening was very dependent on light level. British studies confirm this Canadian finding, the birds becoming active shortly before sunrise, and returning to their roosts at, or soon after, sunset. On their evening return the Merlins may actively hunt small birds roosting in the same area.

In winter it is usually impossible to differentiate between adult female and juvenile male Merlins and consequently many refer only to 'blue' and 'brown' Merlins (see, for example, Fig. 14. The photograph opposite is of an adult Eurasian male, that below of a first winter Eurasian male.

Survival

The age of Merlins at the time of their death can only be ascertained with any certainty for ringed birds, and use of that data has both advantages and disadvantages. The advantage is that age is known with certainty for birds ringed as nestlings. There are several disadvantages, however. The major problem is that 90% of ringed birds are never recovered, so that information for the total population has to be inferred from a very small sample. A second disadvantage is that recovered birds are likely to have died close to, or as a consequence of, human activity, which means that birds dying as a result of 'natural' causes – i.e. not being hit by a vehicle, flying into windows or other unforgiving structures, or being shot – are less likely to be recovered.

In a study of data on recovered Merlins ringed as nestlings in Britain, Heavisides (1987) lists the ages of 242 birds found dead or so sick that they did not recover (Fig. 27). The first block of the histogram of Fig. 27 denotes recoveries of birds found dead by the end of August of the year in which they were hatched. These birds represent 17.4% of the total recoveries. Newton *et al.* (1978) continued to observe Merlin nest site areas during the period in which the juveniles were wholly or partially dependent on their parents for food and noted that the loss rate was 7-15%. The wide range of the estimate was due to the difficulty of knowing with certainty whether a juvenile had been lost as a consequence of starvation, due to not developing adequate hunting techniques at a time when parental feeding had been reduced, as a result of predation, or as a consequence of early dispersion from the nest area. However, the data do suggest that perhaps 10% of Merlin nestlings do not survive until the start of their first winter. In his study of Merlins in Ireland's Wicklow Mountains, McElheron (2005) notes the death of all four near-fledgling in one nest. They had not been predated and there was no clear cause of death, leading to conjecture that they had died during a period of appalling weather when, perhaps, the adults had been unable to hunt and provide food, or when one adult had died and the other had abandoned the nest, either scenario could have led to death by starvation and exposure.

Once the adult birds have stopped feeding their offspring – there is no concrete evidence to indicate that parent Merlins engage in active attempts to expel their brood, but evidence from other falcon species (for instance the Eurasian Kestrel) suggest that parent birds may move their youngsters away from preferred hunting areas – the juveniles must rapidly

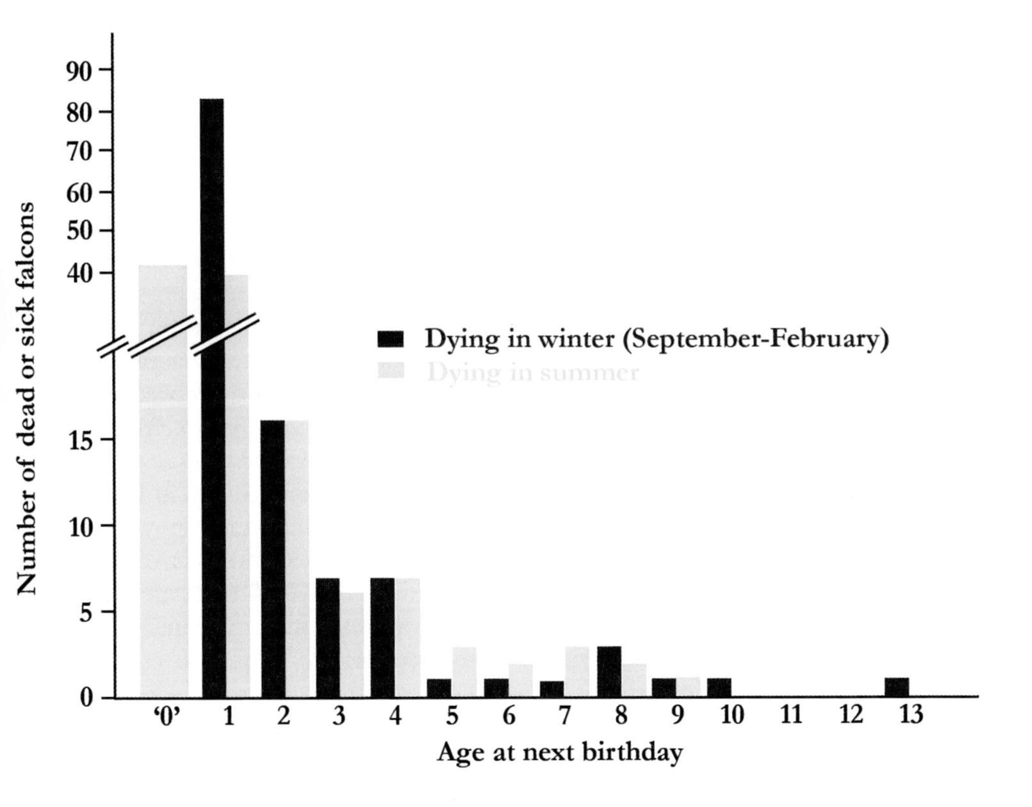

Figure 27
Age distribution of Merlins recovered dead or terminally sick with rings which allow their ages to be determined. In general the sex of the recovered bird was not noted by the finder and no attempt has therefore been to separate the recoveries into male and female.
'0' – dying before August in year of hatch.
Drawn from data in Heavisides (1987).

acquire the hunting skills necessary to feed themselves. As most Merlin populations are migratory, the juveniles must then transfer these skills to a new area, and also cope with weather conditions which are usually poorer, and potentially with a reduced prey population as well. The combination means that many starve, or are unable to fight disease or parasitism because hunger has reduced their overall condition. Fig. 27 reflects this higher death rate in first year birds during the winter, with 82 of 121 (67.8%) of recovered yearlings being found dead in winter. Excluding deaths of fledglings, the deaths of first year birds represent 60.5% (121 of 200) of recovered birds. The first year of a Merlin's life is hazardous, with the first winter being the most hazardous part of it. While the skew towards first year birds makes the calculation of mean life-span difficult, it appears that most Merlins that survive their first winter live 3-5 years, with very few reaching an age in double figures. The data of Heavisides (1987) shows only two birds living to their tenth year, one dying at age 10, the other at age 13: from various studies the latter can be considered to have been a very old small falcon indeed.

A check of the mean life span suggested above is possible from the work of Espie *et al.* (2000) who studied the effect of age on the breeding performance of the urban Merlins of Saskatoon which had been subject of numerous studies (many of which have been mentioned here) over a period of many years. The study of Espie and co-workers has already been mentioned above (see *Chapters 7 and 8*) because of its importance in the understanding of differential breeding in male and female Merlins, and Lifetime Reproductive Success. The study also allowed the age profile of breeding birds in Saskatoon to be established. This is shown in Fig. 28 for 53 female Merlins of known age, and 105 male Merlins of known age. From these data the mean life spans of the two sexes were calculated as 3.15±1.91 years for female Merlins and 2.67±1.51 years for males. The study was also able to estimate the age for other male Merlins, which resulted in a calculated life span for that cohort of 2.58±1.18 years, in good agreement with that for known-age birds. The oldest bird identified in the study was a female aged eight years: the oldest male was aged six years.

Life span was also calculated in a study in northern Sweden by Wiklund (1995) who found a mean life of 2.2±1.4 years for 92 females. In a further analysis of data from his study (Wiklund 1996) calculated a mean life span for 55 male Merlins of 4.0±2.1 years. Each of these life spans is in reasonable agreement with the spans calculated by Espie *et al.* (2000) when the standard deviations are taken into consideration. Wiklund (1995) notes that the oldest female in his study group was aged six years, though one female which was not in his study group bred at age eight years. In Wiklund (1996) the oldest male was aged seven years. In each case these ages are in good agreement with the Canadian study on *F.c. richardsonii* and are comparable with the numbers of older Merlins noted by Heavisides (1987). The data are also consistent with the idea that most Merlins that survive their first winter live 3-5 years. However, as can be noted in Fig. 27, one Merlin is known to have survived into its 13th year. That bird, a male, was ringed in County Down, Northern Ireland, on 21 June 1959, and recovered on 1 January 1972: the bird was recovered only 35km from where it had been ringed (Heavisides 1987). Merlins reaching ages in double figures in the wild are probably extremely rare, though Rydzewski (1974) also mentions a bird reaching the age of 10 years 7 months 20 days. The mean life of 3-5 years is also consistent with the finding of Marsden

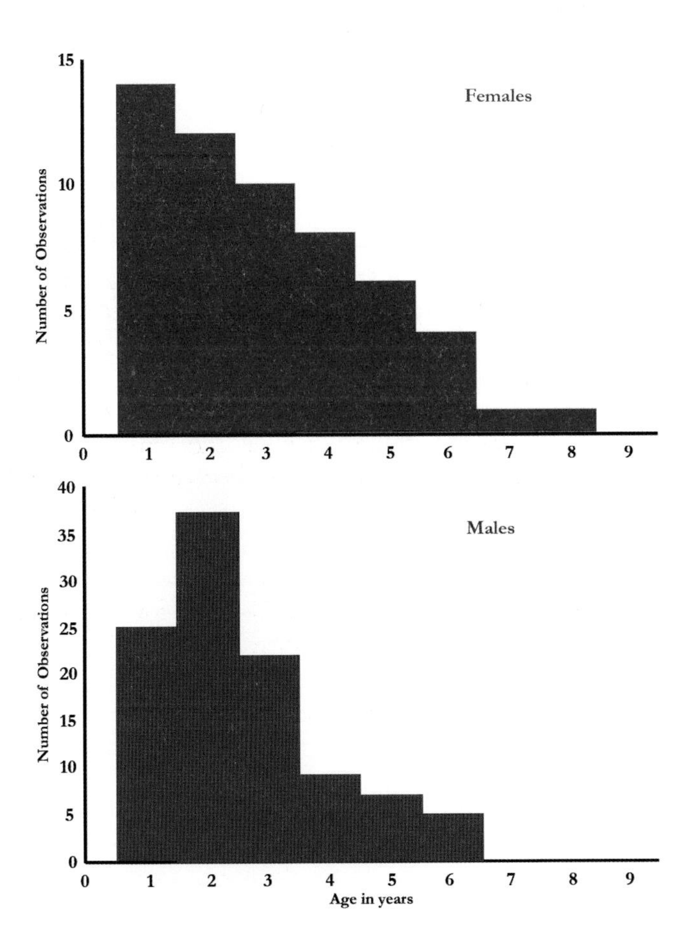

Figure 28
Age distribution of male and female Merlins.
Redrawn from Espie *et al.* (2000).

(2002) in the DNA profiling study of Merlins breeding in north-east Scotland. Marsden found that the typical breeding lifetime of a Merlin was just one or two seasons (giving mean lives of both sexes of 3-4 years), though two females which were at least nine years old and a male of seven were noted.

The study of Lieske *et al.* (2000) on 1354 urban-dwelling Merlins ringed over a 12-year period in Saskatoon, Canada has already been mentioned (see *Breeding Territory and Mate Fidelity* below). As well as providing data on territorial fidelity, the study provided data on the survival rates of adults and juvenile birds. These data suggested that the adult survival rate (male and female combined) was about 62%/year, but that adult survival was

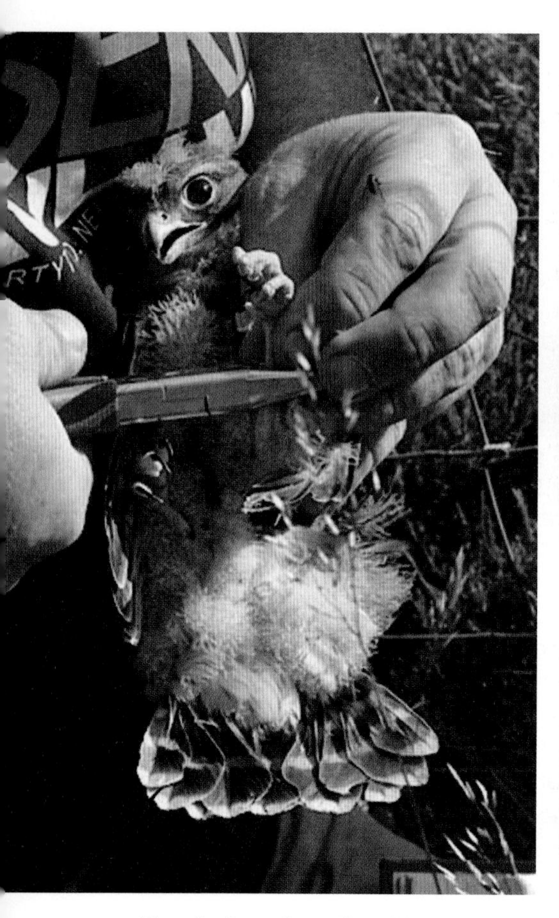

The ringing of nestling Merlins allows information on survival rates, as well as migration to be gathered.
Mike Price.

negatively correlated to population density. So, although the city habitat is excellent in terms of prey availability and relative freedom from nest predation, the rigours imposed by having to defend territories and, in the case of males, to defend nest sites from intruding males seeking to mate with their females, causes stresses which result in reduced life spans. In the study, no bird, male or female, was recaptured more than five years after first ringing (with a median time of three years). The annual survival calculated by Lieske *et al.* (2000) was comparable to an earlier study of the same urban-dwelling Merlin population by James *et al.* (1989), though the latter calculated mortality rather than survival, calculating values of 29% for females, and 31% for males.

In a further study in Arctic Sweden (Wiklund 2001) Wiklund noted that the loss of females during the winter was higher than that of males, assuming that the most likely reason for failure to return was winter mortality. For males winter survival seemed dependent on breeding density – in years of low breeding density winter survival was higher, as it was in years of high rodent population. However, in years of very high rodent population, male winter survival declined, Wiklund tentatively suggesting a proximate cause being the fact that in such years clutch sizes were higher and so more fledglings were produced. Wiklund does not specify a potential exact cause, but the implication is that a higher work rate causes a loss of body condition in the male which reduces the likelihood of survival in winter. For female Merlins, Wiklund found that winter losses were higher than for males, but were independent of both breeding density and rodent population, and speculated that higher dispersal distances post-breeding might result in an inherently lower return rate irrespective of survival rates, or that the high energetic costs of reproduction in any year might result in loss of body condition.

Causes of Death

James *et al.* (1989) looked at the cause of death of 89 recovered Richardson's Merlin in Canada and found 43% had died as a result of collisions, 7% by shooting, 2% from poisoning, 1% as a result of adverse weather, with 44% of unknown causes. In many of the latter, starvation, disease or a combination of the two may have been the cause. The high rate of death through collisions is not surprising in an urban population as it is known (e.g. Hager 2009) that death rates by collisions with windows and vehicles are high for all urban-dwelling raptors: Hager quotes window-strike mortality as the main cause of death for urban-dwelling falcons and owls.

In Iceland Nielsen (1986) notes that of 45 recovered Merlins, 81% were juvenile birds. Five of the deaths (11%) were due to collisions with objects, 12 birds (27%) had been shot, three had been 'injured' (7%) and died of the injuries, three more (7%) were classified as 'exhausted', one (2%) had been 'trapped', while the majority (21 birds – 47%) were 'dead'. Here again, it is likely that exhausted or 'dead' birds were suffering from disease or starvation, while injured birds may well have been involved in collisions which they, initially, survived. It is not at all clear exactly what 'trapped' might refer to.

In his study of the recovery of Merlins ringed in the British Isles from 1911 to 1984, Heavisides (1987) provides data on the cause of death of 254 birds recovered dead or that were sick/injured at recovery but which subsequently died. In this study 37.8% were 'found dead' with no further explanation, 20.9% were killed by man, 9.8% died after being recovered sick or injured and the remaining 31.5% were classified as 'other' a category which included the falcons being road casualties, flying into obstructions, starvation, being killed by a cat, drowning in a water tank and being 'found on a ship'. What is interesting about this 70 year study is the variation in the fraction of recovered Merlins which were of 'killed by man'. This fell from 75% in the period 1911-1920 and 76% in 1921-1930, to 62% (1931-1940), 30% (1941-1950), 27% (1951-1960), then 16% (1961-1970) and around 5% after 1970. It is believed that most of the deaths in the earlier periods were from shooting. Prior to the passing of the 1954 Protection of Birds Act in Britain the shooting of any raptor seen on a game bird estate was routine, the fact that Merlins could not take adult game birds and rarely took game chicks being an irrelevance. Rowan (1921-22 Part II) notes that on one occasion a gamekeeper was watching the area where he knew Merlins were nesting, waiting for the male to return with food so he could see the female rise to meet him and so reveal the nest position. The male Merlin flew in and, on seeing the keeper, dropped his prey in surprise, a grouse chick landing on the keeper's shoulder. As the keeper's intention may have been destruction of the nest, the dropped chick probably mattered little, but would doubtless have reinforced the view that all avian predators were a threat to the grouse.

Heaviside's (1987) data clearly shows the impact of the Act, though, of course, the indiscriminate shooting of birds of prey by gamekeepers still continues, if on a thankfully reduced rate. However, while it would seem that the killing of raptors (Merlins included) by gamekeepers is inevitably bad for population numbers, a study on Hen (Northern) Harrier numbers in one area of Scotland (Baines and Richardson 2103) suggests that the position is not clear cut. On a moor managed for grouse shooting, the effect of a reduction in the shooting of the harriers led to a ten-fold increase in the number of breeding pairs over a five-year period, but the cessation of management (and, therefore, the absence of keepers, led to an increase in the population of crows and Red Foxes and a reversal of the harrier population increase. Gamekeepers can be both the raptor's friend and his foe.

Chemical Contamination

In work on the effect of organochlorine pesticides by Derek Ratcliffe (**Box 27**) and others it was shown that egg thinning and breakage was a significant cause of population decline in the Peregrine Falcon. But the damage was not restricted to the larger falcon, Newton *et al.* (1978) noting the Merlin is equally susceptible to the effects of organochlorine contamination.

The effect of organochlorine contamination on the Merlin population is considered in *Chapter 12*. But while that specific problem has been overcome, it is worth noting that other man-made contaminants, or contaminants which derive from industrial or agricultural processes, are

Box 27
Derek Ratcliffe (1929-2005) was a leading British environmentalist and ornithologist who was appointed to head an official (i.e. government sanctioned) investigation into the status of the Peregrine Falcon in Britain after complaints by owners of homing pigeons that Peregrine numbers were increasing. The investigation, begun in 1961, noted the Peregrine population was actually in steep decline, subsequent work identifying organochlorine contamination of the birds, as a consequence of the falcons consuming prey which was itself consuming insects contaminated by organochlorine pesticides. In a report published in 1970 (Ratcliffe 1970) the link between contamination and eggshell thinning, with consequent egg breakage and a reduction in fledgling falcons, was established. The work ultimately led to the banning of organochlorine based pesticides. For an excellent history of the pesticide story see Ratcliffe (1993).

still widespread in the natural environment and that each may influence the breeding success of the Merlin (and other raptors).

In the latest of a series of reports on studies into environmental pollutants in the eggs of birds of prey in Norway – a work programme which became part of the *Program for Terrestrisk naturovervåking (TOV)* (Programme for Terrestrial Monitoring) in 1992 – Nygård and Polder (2012) used data gathered over the previous 40-50 years to produce long-term trends in pollutants. The study showed that the levels of legacy pollutants continue to decline, and that the majority of eggs show concentrations below the believed critical levels, but that levels of persistent organic compounds (PCBs) have stabilised, rather than continuing to decline, in some species. In the eggs of Merlin, Peregrine Falcon and White-tailed Eagle the by-products of DDT (Dichlorodiphenyltrichloroethane), particularly DDE (Dichlorodiphenyldichloroethylene), and the persistent organic compounds PCB (Dichlorodiphenyldichloroethylene) and HCB (Hexachlorobenzene) are still the dominant pollutants, and DDE, which is known to be a significant egg-shell thinning contaminant, is higher in Merlins than other birds of prey and egg-shell thicknesses have not yet returned to pre-organochlorine usage levels. Fig. 29 (overleaf) shows the variation of some of these persistent contaminants, and also of Mercury in the eggs of Merlins over the period 1975 to 2010. Merlins, together with Peregrines and White-tailed Eagles, also have the highest levels of mercury. More worrying, as it took many years for the effect of organochlorine contamination to be understood, both brominated flame retardants and perfluorinatedalkyl compounds are now being seen in raptor eggs. Levels are small, but little is known about the biological

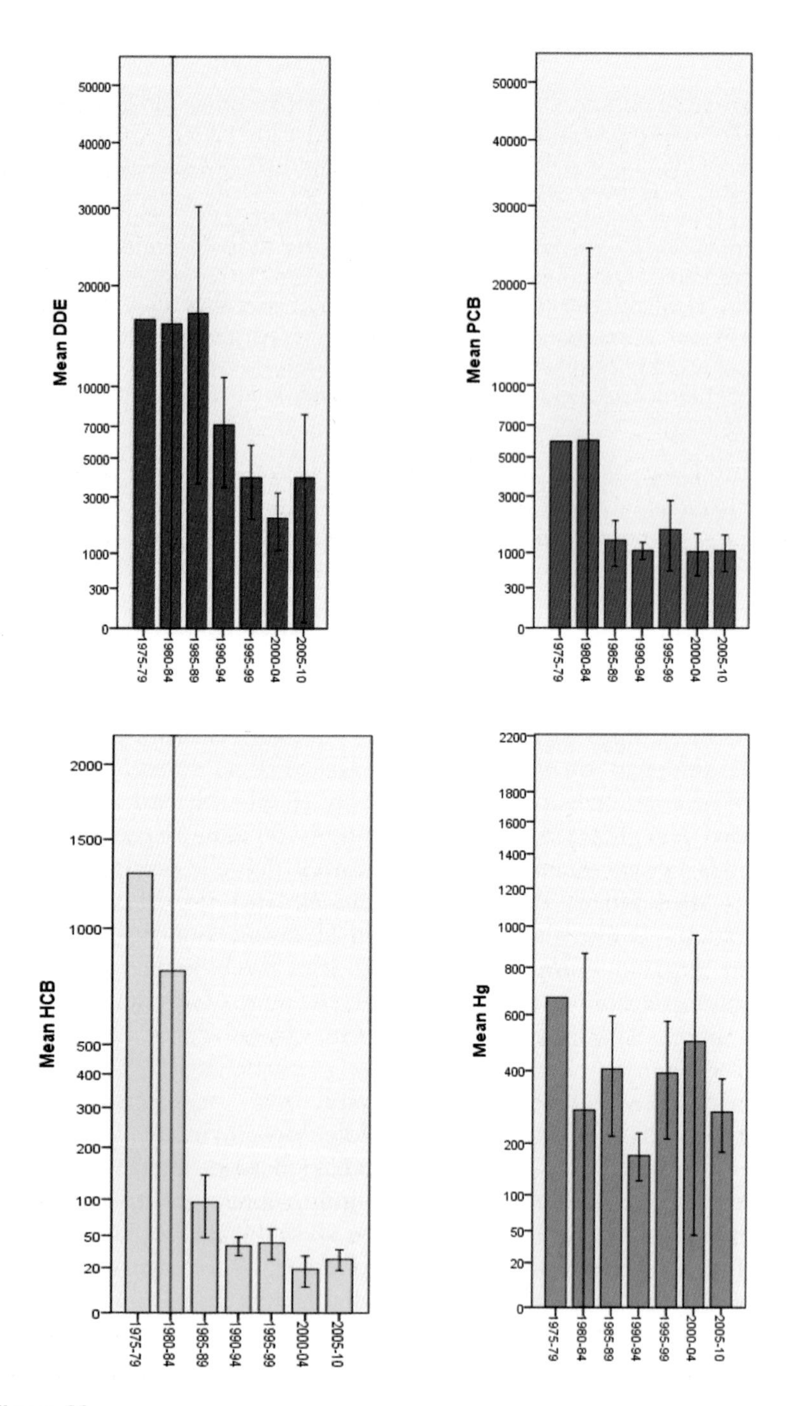

Figure 29

Variation of levels of contaminants in Merlin eggs in (parts/billion fresh weight) for DDE (to be absolutely accurate, p', p' DDE); PCB; HCB; and Mercury.Redrawn from Nygård and Polder (2012).

effects of these compounds, particularly over the longer term. While not specifically mentioned in the work of Nygård and Polder, concerns have lately been expressed over the increased usage of neonicotinoids which are linked with decreases in bee populations, as a result of which they have been banned in several countries, though not, as yet, for instance, in the British Isles and USA. A recent paper (Hallmann *et al.* 2014) noted that in a study in Holland there was a clear negative correlation between the populations of 15 insectivorous bird species and the high surface water concentrations of imidacloprid, the most widely used neonicotinoid, populations declining by an average of 3.5%/year (the effect was seen most markedly in Barn Swallows) in areas where the concentration of imidacloprid was higher than 20 nanograms/litre, even after correcting for land use changes. These declines have been observed only since the introduction of neonicitinoids to the environment in the last decade.

Neonicotinoids are water soluble, persistent in the soil and it is estimated that 95% of the pesticides applied to crops are taken up by the plants, the rest reaching the soil. And while the study of Hallmann and co-workers noted the effect of 20 nanograms/litre of imidacloprid, in places concentrations up to 50 times higher were noted. The Dutch study therefore raises very real fears that in the future, populations of avian predators will begin to show declines. Other studies have also shown that neonicitinoids are lethal to non-target insects at very low doses (e.g. Charpentier *et al.* 2014), leading to fears that, over time, accumulation of the chemicals in species further along the food chain may be significant. Because of time delays, as with the organochlorines, the impact may not become apparent until the effects are profound and long-lasting, as with the recent problem of diclofenac and its effect on Asian vulture species. There are corresponding concerns expressed for the safety of European vultures now the anti-inflammatory has been approved for use in cattle in Italy and Spain.

Another contaminant which has caused concern over many years is lead, with waterfowl in particular ingesting spent lead pellets, surviving being shot, but having lead pellets embedded within them, or not being retrieved by shooters. Raptors subsequently eating lead-carrying birds then build up lead concentrations. Pain *et al.* (1995) measured lead levels in 424 British raptors of 16 species found dead in the 1980s/early 1990s by examining the liver. They found concentrations associated with lead-based mortality (>20ppm dry weight in the liver) in only one (of 25) Peregrines and one (of 50) Common Buzzards, but they recorded elevated levels in other specimens, including one other bird of each species with a concentration of 15-20ppm. 55% of 63 Merlins had non-detectable concentrations, with 6 birds showing concentrations of 6-15ppm dry weight. Pain and co-workers consider it possible that the

Merlins with high lead concentrations could have acquired the metal by consuming waders (Redshank or Common Snipe) which had been shot, but survived. A later study (Clark and Scheuhammer 2003) found a similar result in a study in Canada when 184 dead birds of 16 species were examined. In this study two Golden Eagles, two Red-tailed Hawks, one Bald Eagle and one Great Horned Owl were found to have had >20ppm dry weight in the liver, indicative of lead poisoning. Only one Merlin was examined: it had a liver lead level below 6ppm dry weight. While Merlin are clearly at less risk of acquiring high lead concentrations than larger raptors, as with other contaminants, the long-term effect of a lead burden on species' health and survival is not known. The current regulations on the use of lead shot in North America and Europe have significantly reduced the likely threat in the future. However, Chandler *et al.* (2004) measured the lead levels in House Sparrows in Vermont, specifically to understand the potential effect this might have on Sharp-shinned Hawks and Merlins. Each of the raptors have shown increased populations in urban areas, and while the increase in the hawk population is correlated with increased House Finch numbers, the raptor is known to take House Sparrows as well: urban Merlins almost exclusively feed on House Sparrows. Despite the fact that using lead as an additive to gasoline/petrol has now ceased, Chandler and co-workers found elevated levels in urban House Sparrows and conjecture that this results from the ingestion of grit to aid digestion, lead being a persistent contaminant of the urban environment. The study indicated that in general the level of lead in House Sparrows was low, but that individual sparrows might carry significant burdens. Chandler and co-workers note that a small number of raptor deaths from lead poisoning have been identified in US, and consider that further work is warranted to identify the level of risk to urban-dwelling raptors.

Overall it seems that while pollutants have historically reduced the reproductive success of Merlins, and may still influence that success, at present they are unlikely to be a direct cause of death, though, of course, a contaminated bird may be more likely to succumb to disease or to starvation if its hunting ability is impaired. Identifying the cause of death of recovered Merlins is difficult in most cases as unless the cause is clear, e.g. the bird has flown into an object and sustained fatal injuries, the finders may not have the experience to discern causal effects and most dead Merlins are not subject to autopsy. It is also inevitably the case that most recovered Merlins die close to human habitation (often as a direct result of human activity) because those that die in more remote areas may be subject to post-mortem predation and so are rarely recovered. Nevertheless statistics on recovered Merlins have been accumulated.

Migration

While winter is clearly a time of great stress for Merlins, with the potential for limited prey, reduced hours for hunting, and poor weather, which can both further limit the possibility of hunting as well as bringing cold ambient temperatures and high winds and an increased risk of hypothermia, migration is also stressful. In a study by Klaassen *et al.* (2014) the risks of migration were highlighted. While the species studied by Klaassen and co-workers were Marsh Harrier (*Circus aeruginosus*), Montagu's Harrier (*Circus pygargus*) and Osprey, each of which is a long-distance migrator flying from breeding sites across Europe to wintering sites in north Africa, the general points made by researchers seem appropriate to all migratory species. Klaassen *et al.* (2014) concluded that mortality was six times higher during migration than in the 'stationary' periods which separate the migration flights. They also found that the total death rates in all four periods (i.e. spring and autumn migrations, breeding and wintering seasons) were approximately the same, the higher mortality occurring in the migration periods that were shorter than the stationary periods. As a consequence, half of all deaths in their studied species occurred during migration. Klaassen and co-workers speculate that mortality during the spring migration may be responsible for short-term annual variations in survival and population size, while autumnal migration mortality may regulate long-term population.

Diseases, parasites and pests

Only a single study on the causes of death from disease of Merlins by examination of dead birds has been identified. Cooper and Forbes (1986) examined 35 dead Merlins, all of which were captive birds, and also made clinical assessments of 13 live birds (11 of which were captive). Of the 35 dead birds, no diagnosis could be made for eight of them. Eight others had fatty liver-kidney syndrome, a condition which seems only to affect captive birds and may be associated with diet or in-breeding. Three birds died of enteritis/proventriulitis, three of coccidiosis. Three birds had peritonitis, two died of septicaemia and one of pneumonia. Two died of anaesthetic or surgical shock during treatment. Two birds died as a result of unidentified trauma, two had been poisoned, and one bird had been euthanised. Of the live birds, five had bumblefoot, the remainder having other minor problems.

Following publication of the work of Cooper and Forbes (1986) the first recorded episode of lymphoid leukosis was observed in a three-year old captive male Merlin, one of a pair (Higgins and Hannam 1985). The disease was previously known in poultry, and was probably contracted from feed stuffs fed to the captive Merlins. The Merlin died, and an autopsy revealed a lymphosarcoma in the liver.

Cooper and Forbes (1986) also found nematodes on three of the dead birds, and it is known that Merlins may also be infected by *Trichomoniasis gallinae*, a motile single-celled protozoan which lives in the mouth, throat and crop, and causes 'frounce' or 'canker' which can be fatal. The potential burden of other parasites in Merlins is unknown, though from work on other small falcons, particularly the Eurasian Kestrel, it is known that that species may be infested with other worms, Caryospora spp., trematoda flukes and cestoda tapeworms. Krone (2007) provides details of other protozoan endoparasites which are known to infest raptors. Wright (2005) records finding parasitic flat-flies (Hippoboscidae) on young Merlin chicks, but considers that the birds had rid themselves of them by the time they had fledged. Since such parasites are known to infect all falcon species (Maa (1969) and Matyukhin *et al.* (2012) found the species *Ornithomya chloropis* on Merlins, though it was the only species found) this may not always be the case. While parasites may not kill the falcons they infest, they may weaken the bird and so be a contributory factor to death from other diseases or from starvation.

Body parasites of both adult and young Merlins are also known, these include screw-worm flies, louse flies and Mallophaga chewing bird lice. In a study of raptor nests as a habitat for invertebrates, Philips and Dindal (1977) note that nests are a potential home to parasites which may infect adult birds and nestlings, as well as animal saprovores (invertebrates which feed on carrion, excreta, pellets etc.), and humus fauna (invertebrates which feed on nest material). Philips and Dindal note five species of arthropods from two families which have been found in Merlin nests (though do not specifically name them) as well as listing the insects commonly found in raptor nests. The insects include Hippoboscidae, as mentioned by Wright (2005).

Merlins, particularly nestlings, in nests in the Arctic or sub-Arctic fringe, are also at the mercy of mosquitoes. Any bare skin on a chick can be exploited, and mosquitoes may even find their way through the down feathers to reach the body. Nestlings may also be attacked by Simulium spp. (Black Flies). Ellis (1976) found five Merlin nestlings infested with the flies in central Montana: one week later the nestlings were gone. While predation is the likely cause, it is possible that blood-letting by the flies was the cause of death or a contributory factor to it. Ellis does not record which Black Fly was involved, but from personal experience further north in North America, Black Fly attack can be likened to them using a knife and fork rather than the mosquito's rapier-like syringe mechanism, leaving open wounds which may become infected.

Mallophaga on an Icelandic Merlin.
Jóhann Óli Hilmarsson.

Human Interference

One Interesting aspect of Merlin studies in North America was the realisation that human studies of the species might actually be causing deaths or reducing survival of both adults and nestlings, a valid concern being raised by Oliphant (1974) that climbing trees, particularly during laying or incubation, might cause nest desertion. Indeed, even if desertion did not occur, concerns over disturbance of birds, particularly nesting birds, are valid, not only in terms of bird welfare and the ethical issues which that raises, but in terms of the validity of the research if disturbance results in non-natural behaviour. However, Rosenfield *et al.* (1987) suggest that provided the frequency of nest visits is minimal, that every care is taken to avoid startling incubating or brooding females (who might then fly off quickly and dislodge eggs or nestlings which fall to the ground and are lost), and provided visits at critical times, e.g. during periods of low ambient temperatures when the absence of the female might cause eggs or chicks, who are unable to thermoregulate, to chill, then evidence suggests there is no effect on breeding success.

As previously noted - see Box 12, *Chapter 5* - a study of the radio-tagging of urban-dwelling Merlins in Saskatoon has showed that this had no discernible effect on the falcons. A further study by Holmes *et al.* (1993) looked at the effect of vehicles and walking humans on the behaviour of wintering raptors on grassland in Colorado. They found that in general the raptors were less concerned by vehicles, tending to flush greater distances at the approach of a human than for an approaching vehicle. Also in general, larger raptors flushed further than smaller ones (e.g. Golden Eagles would travel about five times further than American Kestrels) which the researchers consider reflects the great level of persecution larger raptors have been subjected to (though it would be imagined that the size of the bird would also relate to distance travelled once airborne because of inertia, and the longer time needed to glide to perch). Interestingly, for smaller raptors, particularly the Merlin and American Kestrel the flush distance for vehicles and humans was very similar. Holmes and co-workers conclude that researchers should self-impose 'buffer' zones to minimise flushing, and so minimise disturbance. For Merlins they recommend 125m as a distance which would reduce the likelihood of flushing by 90%.

Ruddock and Whitfield (2007) also calculated the approach distances before disturbance for Merlins (and other sensitive species of the Scottish avifauna), but during the breeding season rather than at a wintering site. Ruddock and Whitfield defined 'static disturbance' and a corresponding 'alert distance' (AD); and 'active disturbance' and a corresponding 'flight initiation distance' (FID), the former involving increased vigilance by the bird or an alarm call, while the latter induced a flight response, either towards or away from the observer. The researchers also considered two breeding phases, incubation and chick rearing. They then contacted researchers of the various species and spoke to them about their experiences with birds, and used these conversations to build up ranges for their two parameters and so were able to define median distances for each. For Merlins the median AD during incubation was 225m (range <10m to 500m – 80% values, arrived at by taking the upper and lower 10% values from their experts and discarding them), while the incubation FID was 30m (80% range <10m to 300m). For the chick rearing phase the corresponding values were AD – median 400m, 80% range 10-500m – and for FID – 225m, 80% range 10-500m.

The flight and tail feathers of this Eurasian Merlin are in excellent condition as it nears fledging.

Friends and Foes

While bird species often have foes, the suggestion of them having friends would raise little support among expert ornithologists. But here we stretch the concept of friendship to include some species which have interesting relationships with Merlins.

Friends

Co-operative behaviour between Merlins has already been considered, the species roosting communally during winter (see *Chapter 9*), and co-operating to hunt (see *Chapter 5*). There are also instances in which unrelated, non-breeding adults (usually first-year males) will aid in both nest defence and food transfers to fledglings (see *Chapter 7: Copulation*). Dickson (1988) also reports a Merlin close to hunting Hen (Northern) Harriers and suggests that the association was mutually advantageous in terms of locating prey.

While it can hardly be suggested as friendship, Newton *et al.* (1978) found a level of tolerance between Merlins and Eurasian Kestrels, almost certainly due to the fact that they were not in competition for prey, being primarily avian and mammal feeders respectively. In some cases in their study area Newton and co-workers found that the two species would use particular nest sites in consecutive years and would also occasionally nest on the same cliff within 30m of each other, the Kestrels using nest sites set higher on the face, the Merlins lower ones. The present author also found a site in Scotland at which a Eurasian Kestrel pair was occupying a nest box less than 30m from a Merlin pair which was also breeding in an artificial nest. However, despite this tolerance and close association, attacks by Merlin on Kestrels have been recorded.

Suggesting considerably more than mere tolerance, in Scandinavia Fieldfares may choose to nest close to Merlins, and while it would be assumed that this was in order to gain the advantage of the falcon's aggression towards corvids, the situation is much more complex and interesting than that. Fieldfares are communal nesters and Slagsvold (1979), studying the thrushes near Trondheim, Norway, noted a significant difference between predation of Chaffinch nests for those finches which nested close to, rather than away from, Fieldfare colonies, but also in the clutch sizes of Chaffinches and Bramblings which nested close to the colonies. In each case the difference was stark. Of six Chaffinch nests far from a Fieldfare colony five were predated, chicks being reared in only

The relationship between Hooded Crows (left), Fieldfares (right) and Merlins is both complex and ambiguous. The crows are the chief nest provider for the falcons, but predate Merlin eggs and nestlings. Both Fieldfares and Merlins provide some defence against marauding crows, but Merlins may predate Fieldfare broods.

one, while of 25 nests in a colony only three were predated, chicks being reared in the remaining 22. In terms of clutch sizes, Chaffinches normally laid a clutch of 4.71 eggs (mean of 318 nests) compared to 4.69 eggs for 29 clutches in Fieldfare colonies, while Brambling clutches rose from 5.83 eggs (N=53) to 6.92 eggs (N=12) for those in Fieldfare colonies. Slagsvold noted that Chaffinches took 3-4 days longer to build nests than the Fieldfares, and laid, on average, a day earlier. Bramblings, which arrived later than the other two species, could therefore locate Fieldfare colonies that were already established and also select nest sites that took advantage of the siting of colony nests. As a consequence the Bramblings laid larger clutches and raised larger broods. Slagsvold noticed that other birds were also taking advantage of Fieldfare colonies – Common Redpoll, Icterine Warbler (*Hippolais icterina*), several species of shrike (Lanius spp.) and the local small falcons. One of the latter was the Merlin.

Hogstad (1981) supplied the answer to the question of what benefit the species nesting close to Fieldfare colonies were achieving. Fieldfares have developed communal nest defence strategies, which not only include mobbing of potential predators, but 'defecation attacks' in which the mobbing birds eject fecal sprays which soil the plumage of avian predators, and have been known to injure, and even kill, raptors and owls. In a study in a sub-alpine birch forest in central Norway from 1978 to 1981 Hogstad noted that 33 pairs of Fieldfares nesting close to breeding Merlins (in four colonies, average size 8.25 nests/colony) had higher reproductive success rates than 65 pairs (in 11 colonies, average size 5.91 nests/colony) which were not associated with Merlins, fledging an average of 1.73 fledglings/nest against 0.55 fledglings/nest. Of the 65 nests away from Merlins 85% were completely emptied (of eggs or nestlings) by predators, the remaining 15% being partially predated: of the 33 nests near Merlins 22% were completely emptied, the remaining 78% being partially predated. These figures call into question the suggestion that Fieldfare nest defence is so good that other species may benefit from it, but, of course, it must be borne in mind that the predation rate for all open nest passerines is very high so any defence which offers a marginal gain is of value, and Hogstad notes that 0.55 fledglings/nest is lower than the mean value of 0.8 during the ten years of his study. It is also worth noting that in a separate study of Fieldfares in Sweden, Wiklund and Andersson (1980) found that in non-colony Fieldfare nests the mean number of fledglings was 0.72±0.34, while in colonial nests was 1.69±0.33: predation rates are high in all Fieldfare nests, but are significantly higher in nests away from colonies.

Hogstad also suggests an answer to another obvious question – if nesting close to Merlins is so advantageous why are Fieldfare colonies not even larger? Hogstad suggests that as Fieldfares nest preferentially close to suitable feeding grounds, colonies can only grow to a limit imposed by intraspecific competition for food.

Since, in general, Merlins breed earlier than Fieldfares, it would seem that the main benefit here is to the thrushes since they have a choice of nesting close to, or far from, the falcons. In a study in Arctic Sweden, Wiklund (1982) noted that Hooded Crows were the main predator of Fieldfare nests, mainly taking eggs. Merlins were extremely aggressive towards the crows which were also a major predator of falcon eggs (not only in Sweden, but elsewhere in Europe, e.g. McElheron 2005 in Ireland who notes that Merlins will not tolerate the crows near their nests): the Fieldfares clearly benefitted from this aggressive behaviour, the pairs at the centre of the colony benefitting most because for them colonial thrush nest defence was also most rewarding. But it is not only the Fieldfares that are benefiting, because, as Wiklund (1979) notes, the Merlins also benefit from the arrangement with increased breeding success, a study in northern Sweden finding that 70 of 91 clutches laid by Merlins within a Fieldfare colony produced young, while only 13 of 34 nests outside colonies did so, a difference which is statistically significant at the 99.9% level.

Since Fieldfares and the other species which choose to nest close to their colonies are potential prey for the Merlin, the former clearly consider it worth the risk of predation to enhance their chances of breeding success, making the relationship between the Merlin, the Fieldfare and the other passerines a fascinating example of commensalism. Ivanovski (2012) has also suggested, from studies in Belarus, that a similar relationship may also exist between Wood Pigeons (*Columba palumbus*) and Merlins, having found several instances of the nests of the two species within 30-50m of each other and one occasion when a pigeon nest was only 1.5m below an artificial Merlin nest. Such an association is known to exist between Wood Pigeons and Eurasian Hobbies, giving credence to the idea.

However, other studies suggest that not all relationships between Merlins and their potential passerine prey are as fruitful for the smaller species. Working with the urban-breeding Merlins of Saskatoon, Canada, Sodhi *et al.* (1990) counted the number of passerines at distances of 50m to 450m (at 50m intervals) from three Merlin nests. The results indicated fewer birds within 100m than further out, so a revised count was instigated, with 11 Merlin nests, a reduced number of distance intervals, and observed passerines being divided into three categories by weight – species weighing <30g (potential prey), species 31-150g (possible prey) and species >151g (non-prey). The results are indicated in Fig. 30 (overleaf). What is interesting from Fig. 30 is that, as already noted (see *Chapter 5*), Merlins rarely hunt close to their nests, so something more complex than the intuitively obvious suggestion that the reason for the decline in passerine abundance is related to potential predation, is involved, particularly as the decrease in abundance is seen in all species, not only potential prey. Sodhi and co-workers suggest that predation is a factor, predator avoidance being essentially 'hard-wired' into all species which may be predated, but that the species may also alter their behaviour, so that some of the observed reduction in counts of small passerines is simply that they were there, but not seen. For the non-prey species, which include species such as crows, which are harassed by Merlins if they trespass near nests, it is fear of attack, with the potential for injury and the energy cost of pursuit, which makes them avoid falcon nests. Once the Merlins had left their nest, the effect of distance from it disappeared, eliminating the possibility that the differences in habitat were the reason for the observed decline. For the remaining query: why does the passerine count not fall to zero close the Merlin nest? Sodhi and co-workers suggest that, as with the Fieldfares and finches noted in Scandinavia, perhaps for some pairs the risk of predation is outweighed by the protection from nest predation by crows driven away by the Merlins.

The studies in Saskatoon also revealed other behavioural changes in prey species. Sodhi (1991b) found that small flocks of House Sparrows

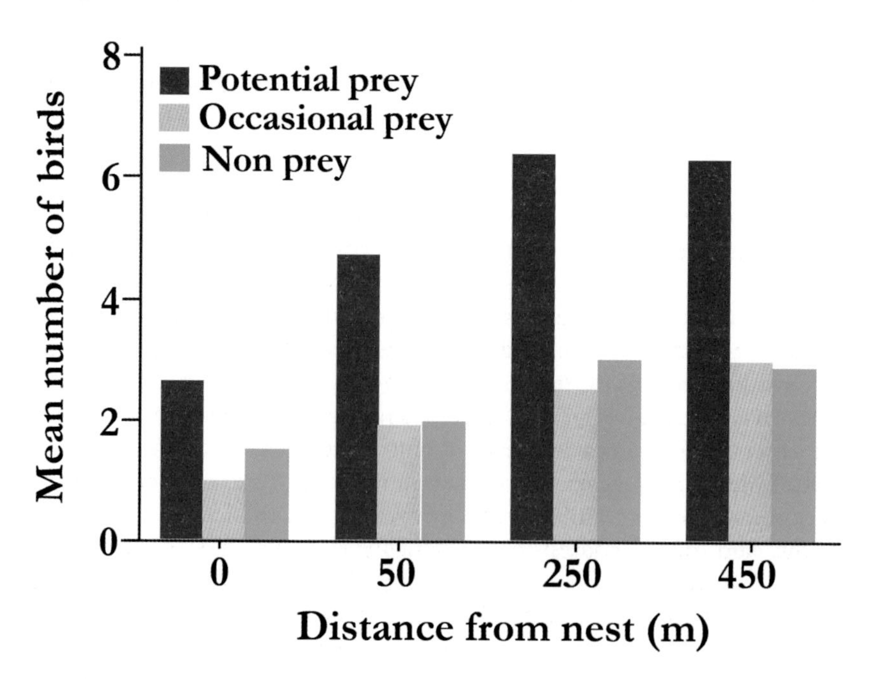

Figure 30
Mean number of birds in three weight categories as a function of distance from a Merlin nest.
Redrawn from Sodhi *et al.* (1990) © Canadian Science Publishing or its licensors.

(which the researcher defined as 1-5 individuals) showed no variation in occurrence with distance from a Merlin nest, but larger flocks (>5 individuals) were less likely to occur at any distance, but were notably absent at distances within 50m of the Merlin nest. Sodhi considers that the absence of large flocks, particularly close to the falcon nest, is an anti-predation strategy, for although flocking is itself a predation defence (as the probability of being selected for any individual is, all other things being equal, lower if that individual is surrounded by many others), flocks are easier for a predator to detect as, self evidently, larger flocks are much easier to spot than smaller ones. In a separate study in Saskatoon Sodhi (1991c) watched the behaviour of three passerine species – House Sparrow, a major prey of the Merlin (58% of diet at the time of study); American Robin, an occasional prey (6%); and Chipping Sparrow, which is rarely predated by the falcon (1%) – when they were close to a Merlin. What was interesting in Sodhi's work was that while both House Sparrows and Chipping Sparrows were significantly more likely to be in concealed positions close to the nest (in Sodhi's study the level of concealment increased within about 100m of the nest), American Robins were not. Clearly the sparrows were using concealment to avoid predation, but the lack of any apparent attempt at increased concealment by the American Robin is puzzling. Sodhi wonders whether the fierce nest defence strategy of the species explains the anomaly, but considers that further work is required before any final conclusion can be reached.

Foes

Merlins are very aggressive to other birds which trespass onto their nesting territories, not only other species, but conspecifics, during the breeding season. As noted above (see *Chapter 7: Territory*) Merlins will also attack conspecifics in winter. When trespassers approach the nest, the male Merlin is usually in a position to give the first alarm call, and will give chase. The female will either sit tight or join in, depending on the age of the nestlings.

Throughout the Merlin's range, corvids (**Box 28** overleaf) are chased, as are other raptors and owls. In a study in Alaska's Denali National Park, Laing (1985) recorded Merlins chasing Golden Eagles and Gyrfalcons, as well as Hen (Northern) Harriers. Laing also observed juvenile Merlins chasing, and being chased by, American Kestrels and Sharp-shinned Hawks. In Britain Wright (2005) noted Merlin attacks on both Peregrines and Goshawks, in each case the Merlin stooping on the larger raptor and hitting it hard. In the case of one female Goshawk, the larger bird rolled on to her back and attempted to catch the male Merlin in her talons, but was out-manoeuvred, and subsequently left the area at speed. British Merlins also have been recorded attacking Short-eared Owls (*Asio flammeus*), while Dickson (1988) records an attack on a Hen (Northern) Harrier (though as noted above, Dickson also records co-operative hunting by the two raptors) and Ingram (1920) notes a Merlin chasing a Peregrine. Also in Britain, Wallen (1992) noted a series of attacks by a male Merlin on a Peregrine which ultimately resulted in the pair locking talons and tumbling towards the ground, separating only about 8m before hitting. Other attacks on large non-raptorial species have also been observed: in northern Ontario, Lawrence (1949) observed an attack on a Great Blue Heron (*Ardea herodias*), a bird which stands 1.2m tall and weighs around 25kg. Merlins have also been seen attacking large gulls: Lusby *et al.* (2011), in Ireland, record an attack on a Great Black-backed Gull (*Larus marinus*), birds which are formidably armed. Millais (1892) also records a Merlin attacking a male Black Grouse (*Tetrao tetrix*) in flight, knocking it down: the Merlin then rose, circled, and flew off. Morozov *et al.* (2013) note Steppe Merlins being particularly aggressive towards Montagu's Harrier, perhaps as a result of the harrier's enthusiasm for taking eggs and young from ground-nesting species.

Merlins may be mobbed by smaller birds, and also suffer predation. Corvids will predate Merlin nests and in his study in Iceland, Nielsen (1986) found that Merlins sharing territories with Ravens were less likely to produce fledglings than those falcons which were not – 74% success with Ravens present, 82% with no Ravens, across a five year study period. Nielsen also found adult Merlins in prey remains at Raven nests, but was unable to be sure whether the Merlins had been predated or scavenged. Other raptors will also predate Merlin nests.

Box 28
As well as defending their nest site against corvids as potential nest predators, Merlins have been observed to chase them in the way that corvids will, for instance, mob buzzards. Merlin pairs have been observed to chase corvids after the breeding season, but as there is little suggestion of mate fidelity this does not appear to be in order to reinforce pair bonding. Brewster (1925) recording observations at Lake Umbagog, Maine, from a diary covering the period 1881-1889, notes an incident where, having been thwarted in around 20 attempts to seize a victim from a flock of 'titlarks' (Meadow Pipits), 'the mortification and disgust of the Pigeon Hawk, because of his ignominious failure...(he) flew listlessly across the marsh to a distant tree. Perhaps it was because of such lost self-respect and with some thought of thereby restoring it that he began harrying an unoffending crow not long afterwards, just as a human bully may correspondingly behave when similarly humiliated'. While today such anthropomorphism is frowned upon, Brewster's account does paint a delightful scene.

Adult Merlins are also predated. In North America (Palmer 1988) Peregrine Falcons, Great Horned Owls, Cooper's Hawk (*Accipiter cooperii*) and Red-tailed Hawk (*Buteo jamaicensis*) are known to take Merlins (Palmer specifically mentions Great Horned Owls killing Merlins in territorial disputes, though the falcon was not then eaten), and it is possible that other larger raptors whose ranges overlap those of the Merlins may also predate them. However, in Iceland, the Gyrfalcon does not appear to predate the smaller bird. The present author, while studying Gyrfalcons in Iceland, found a Merlin nest within 200m of nesting Gyrfalcons and saw no aggressive encounters between the two. Woodin (1980) also noted Gyrfalcons and Merlins nesting in close proximity in Iceland: Woodin saw aggression from the Merlins to the Gyrs, but not the reverse. It would seem that the manoeuvrability of the smaller falcon deters the larger species from attempting to hunt it, though opportunistic attacks are possible and may have occurred. In an examination of 45,000 individual prey items in Merlin nests in a study lasting more than 30 years, the remains of only one Merlin have been discovered (Ó.K. Nielsen, pers. comm.). Lindberg (1983) also records finding a Merlin in Gyrfalcon prey remains in Sweden: the remains constituted just one out of 1869 recovered items, representing 0.18% by number and 0.06% by biomass. These data suggest such attacks by Gyrfalcons are opportunistic and rare.

In Britain Peregrines are known to predate Merlins (Ratcliffe 1993) and it is likely that Golden Eagles, Tawny Owls and Eurasian Sparrowhawks do as well, with varying degrees of success. Roberts and Jones (1999) also consider that the presence of Peregrines in an area historically favoured by Merlin can cause displacement of the smaller falcon even if direct predation is not involved. Working on a seismic research vessel to the west of the Shetland Isles, Gricks (2005) noted the arrival of two Merlins and a Short-eared Owl in late September (the Merlins having perhaps arrived from Iceland). Over the next few days the Merlins killed and ate about 10 Meadow Pipits, but then one of them disappeared. The same day the owl was seen to be eating the remaining Merlin which it had presumably killed. Rebecca (1991) also noted predation by a Short-eared Owl, a dead female Merlin being found close to her nest. Owl feathers were scattered close by and the falcon had owl feathers in her talons. Egg shell fragments were stuck to the Merlin's brood patch suggesting that she had been attacked while incubating.

In a study in northern England, Petty *et al.* (2003) found that 5.4% of the raptors taken by Goshawks were Merlins – the raptor fraction of the Goshawk diet was 4.5% by frequency, 2.5% by biomass: the remains of 13 Merlins were found in a total of 5445 prey items.

Similar Merlin predation is likely to occur in Fennoscandia, and data on owl diets by Mikkola (1983) noted that both Eagle Owls and Tawny Owls predated Merlins: Merlins represented 5 of 705 prey items in the diet of the larger owl, 1 of 33 prey items for the smaller owl. Wiklund (1990b) observed a Rough-legged Buzzard take all three Merlin chicks from a nest over a period of a couple of hours, but considered that poor nest defence by the female Merlin contributed to the loss.

It is likely that similar predation of Merlins by other raptors and owls occurs across the remainder of its Eurasian range. Tømmeraas (1993b) records the killing of a female Merlin by Golden Eagles in the Alta-Kautokeino region of Finnmark in northern Norway. A Merlin pair was nesting about 200m from the Golden Eagle pair and regularly mobbed the eagles when they left their eyrie. One day the female Merlin was observed stooping on an eagle. After one stoop the Merlin gained height in order to stoop again: as it reached the top of its climb and effectively became momentarily motionless as it turned to attack, the second eagle grabbed it from its blind-side. Both eagles then landed on top of the nesting cliff. When the eagles took off again the 'catcher' was still holding the dead Merlin. The eagle dropped the falcon, then swooped down to catch it. It repeated this several times before finally allowing the Merlin to fall into the forest at the base of the cliff. At a ledge on the cliff a juvenile eagle was perched. Tømmeraas considers that the killing of the Merlin was an example of co-operative hunting, the first eagle acting as bait for the falcon with the second eagle waiting its moment to strike. Interestingly Tømmeraas also notes that juvenile Golden Eagles had also been seen to drop and catch objects as a way of enhancing hunting techniques, but in this case the bird dropping the falcon was a mature adult. Although

Tømmeraas does not explicitly state that the dropping of the Merlin seemed aimed at enticing a perched juvenile to attempt a catch, that is what is implied by the report: if that is correct then it seems that when the parent bird realised the juvenile was not interested it merely dropped the dead falcon into the forest and went about its business.

Morozov *et al.* (2013) record an incubating female Merlin being predated by a Hen (Northern) Harrier, while Karyakin and Nikolenko (2009) record predation by both Goshawk and Saker Falcon in the Altai-Sayan Mountains in south-central Russia. Interestingly, Morozov *et al.* (2013) suggest that the Eurasian Hobby may also be a threat to Merlins, based on the noted absence of Merlins near Hobby nests (and vice versa) in areas where the ranges of the two overlap. Two Merlin fledglings were also found with wounds suggestive of a Hobby attack.

Merlins are also predated by mammalian hunters, though in general these would be more likely to take a sick or injured adult, or perhaps an incubating adult which decided to sit tight on the nest, though eggs and nestlings can be taken by both ground-based and arboreal mammals. In Scotland, Pine Martens (*Martes martes*) may be a threat while Morozov *et al.* (2013) note that Merlins may also fall victim to Sable (*Martes zibellina*) in the taiga and to the Steppe Polecat (*Mustela eversmanni*).

Brewster (1925) tells of a Merlin shot by a companion which fell to the ground but was not searched for until sometime later. Blood spattered snow suggested the Merlin had survived the shot and climbed a tree stub, but had then been taken by a mammal which, from the paw prints in the snow was a North American Red Squirrel (*Tamiasciurus hudsonicus*). MacIntrye (1936) records finding an Adder (*Vipera berus*) in a Merlin nest he had gone to inspect. He killed it and then found that three of the four eggs he knew had been laid were missing. He dissected the snake and found the contents of the three eggs. Records of Adder egg predation are rare, but Roberts and Green (1983) note that at a nest in their study area in north Wales one nestling was predated by an Adder: Haffield (2012) reports a similar occurrence in south Wales, as does Shaw (1994) in south-west Scotland. While such snake attacks are likely to be confined to ground-nesting Merlins in Britain and Eurasia, and are likely to be infrequent in North America where Merlins are mostly tree-nesters, DeGregorio et al. (2014) noted that 26% of recorded predation attacks were from snakes. While these were mainly confined to southern latitudes, they were also recoded in Canada, and some of the snake species involved were (at least partially) arboreal, so it is conceivable that snake predation of Merlin nests may occur.

In addition to predation, Merlins are also subject to kleptoparasitism. In a study in western Washington State during winters from 1979 to 1987 Buchanan (1988) observed Merlins feeding on Dunlins, watching a total of 27 successful attacks on the waders. Following five of these attacks (18.5%) the Merlin was subject to a piratical attack. There were two successful aerial pirate attacks, one by a Red-tailed Hawk and one by another Merlin. In one attack a Merlin feeding on its captured prey after landing on the ground was forced to abandon it after suffering an attack by a Hen (Northern) Harrier. In the other two attacks the successful Merlin was forced to drop its prey by another Merlin. In one of these cases the captured Dunlin flew to safety: in the other the Dunlin was taken by a Glaucous-winged Gull (*Larus glaucescens*). On three other occasions a Merlin knocked or forced a Dunlin into the sea, but did not capture it: in each case the injured Dunlin was taken by another bird, two by Glaucous-winged Gulls, one by a Bald Eagle. Buchanan noted that because of the possibility of piratical attack the Merlins were changing their hunting behaviour. The primary attacks by an untroubled Merlin were either surprise on a single Dunlin or a high flight followed by a stoop into a flock of the waders. In each case the success of hunts was lower, declining from 19% to 14.7% for surprise attacks, and from 25.5% to 15.4% for high flight attacks, though the small sample size meant that in neither case was the decline statistically significant.

There was no significant change in the duration of stealth flights (which lead to surprise attacks), but the duration of high flights altered significantly (Fig. 31 overleaf). When raptors were present, all successful high flights lasted less than two minutes, indicating that the Merlins were looking to attack quickly and retreat to safety quickly. When raptors were absent, 41% of successful high flights lasted more than two minutes, some as long as eight minutes: the possibility of kleptoparasitism was clearly influencing the hunting behaviour of the Merlins. In a later study, Buchanan (2009) reported that observations of Merlins at one site had reduced by a factor of 7 in the years covered by his study. While the number of Red-tailed Hawks had decreased, the number of Bald Eagles had increased, but Buchanan considered it was unlikely to be kleptoparasitism by this species which had reduced Merlin numbers as he had never observed such piracy (though this has actually been observed – Dekker 2003). Buchanan (2009) considered, rather, that the reason for the decline in Merlins was the increase in Peregrine numbers as the larger falcons are known to both pirate (e.g. Dekker 2003), and predate, Merlins (eg. Ratcliffe 1993). In a second paper Buchanan (2012) noted that not only had the number of Merlins reduced, but they hunted less often, had shorter hunting flights during which they made fewer attacks on prey, and hunted closer to vegetation which would reduce the chances of predation. Buchanan concludes that the Merlin behaviour shows distinct alteration in line with intraguild predation (IGP) theory which suggests that mesopredators will reduce the risk of IGP if they are sharing habitat and prey with a larger predator.

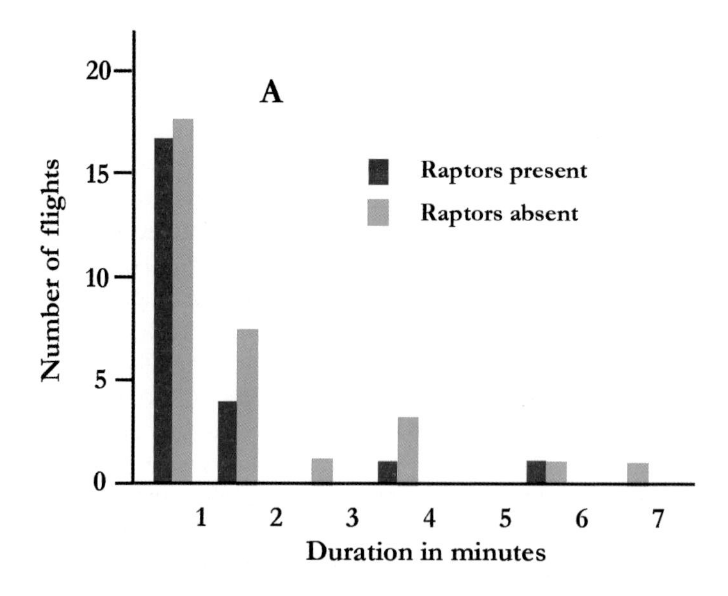

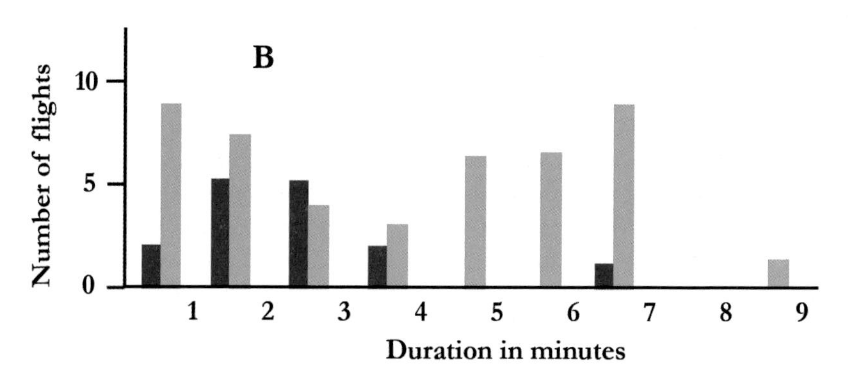

Figure 31
Duration of (A) surprise and (B) high flight attacks by hunting Merlins in the presence and absence of other raptors. Buchanan noted only the presence of raptors in compiling these data: Glaucous-winged Gulls were present at all times. Redrawn from Buchanan (1988).

If humans approach a Merlin nest, the birds may attack. In a study of the behaviour of 19 North American diurnal raptors exposed to the presence of a potential predator, Morrison *et al.* (2006) place the Merlin in Category 4 (out of 4) implying 'very aggressive' nest defence. The categorisation was against non-human predators, Morrison and co-workers considering that against humans all the raptors showed less aggressive behaviour, either because they did not immediately recognise the human as a potential predator, or because the large size of human intruders implied a high risk of injury in any attack. But in Minnesota, Craighead and Craighead (1940) noted that during the building of a hide close to one nest 'both birds attacked us, striking repeatedly and scolding fiercely'. On a subsequent visit both birds attacked again, the male striking and drawing blood. Lawrence (1949), in northern Ontario, also notes such behaviour, remarking that female Merlins are usually the more aggressive of a pair. However, in urban environments in North America, habituation causes Merlins to become much more tolerant of humans, paying little attention unless the human begins to climb the nest tree.

The situation also seems different in the ground nesting Merlins of Britain where an incubating female will often leave the nest silently, possibly with the hope of not drawing the attention of the intruder to the site. The female may then circle overhead, where she may be joined by the male. In such cases the birds may call, but may also remain silent. However, Karyakin and Nikolenko (2009) noted examples of Merlins leading humans away from their nests, and even imitating the injured wing tactic seen, for instance, in waders. Karyakin and Nikolenko note that this behaviour pattern is also seen in the Urals where Merlins attempt to lead men and dogs away from their nests.

In addition to the taking of eggs and deliberate destruction of nests, humans may also be responsible for unwitting disturbance. In Scotland the present author is aware of one Merlin pair deserting a nest when walkers camped beside the tree in which it was set. In addition to the obvious disturbance of the campers themselves, a fire was lit which probably engulfed the nest in smoke. It is not known if the falcons attempted to breed elsewhere. Merlin are also illegally poached in Russia, being collected during expeditions aimed primarily at gathering wild Gyrfalcons for sale in the West. Lobkov (2012) notes that among a 'cargo' of 58 falcons seized on a vessel near Kamchatka in late 2012 there were two Merlins and two Peregrines, the rest of the birds being Gyrs: two other falcons were seen to have been thrown into the sea before the vessel was boarded by the Federal Security Services. At the same time a further 20 Gyrfalcons were found on another vessel and 14 Gyrs were seized at Vladivostok airport. All these birds had been captured in Kamchatka. Whether Merlins had been deliberately targeted or had been taken by mistake (just another falcon as it were) is not clear. Also unclear is whether other Merlins are killed rather than released when taken during such operations after it has been realised they are not as required. Sadly, adequate resources are not available to control the illegal poaching of falcons in Russia.

Population

Nowhere, throughout their range, are Merlins common, but despite this the species is considered of 'least concern' in conservation terms, because of the vast area of the northern hemisphere in which they are found. Best evidence in North America implies that the Merlin population is increasing, while across much of the rest of the species' range, the population is considered to be either stable or slowly increasing. However, the difficulty of establishing Merlin populations with accuracy, because of the remoteness of much of the preferred habitat, means that while the position appears encouraging, at least in the short and medium term, it is open to question.

While present population evidence is reassuring, it is worth recalling that it was not always this way, the consequence of organochlorine contamination was a decline in Merlin population across its range, though principally in the Nearctic and Europe, as there is much less agricultural land in the Merlin's Asian range.

North America

In North America, the decline in Peregrine Falcon numbers prompted researchers to investigate the potential effect of contamination on the Merlin population, particularly as it was clear that the population of the smaller falcon was declining sharply. Fox (1971) noted that in a study area for Richardson's Merlin in Saskatchewan, clutches were disappearing and reproductive success was falling, these observations being followed by the complete disappearance of the species from the study area. As a consequence, Fox looked at collections of Merlin eggs held in museums to assess changes in egg weight over time. He found that while shell weights had been static at about 1.7g during the period 1890-1949, in the period 1950-1969 the mean weight had fallen to 1.3g. This decrease in mean egg weight was accompanied by a reduction in breeding success for Merlins in areas where organochlorine contamination was most likely. At around the same time, Temple (1972c) investigated eggshell thickness in five Merlin (*F.c. columbarius*) nests on Canada's Newfoundland. For five unhatched eggs Temple calculated eggshell thicknesses using the Ratcliffe Index (**Box 29**), then sampled the eggs for the DDT residue DDE, and also sampled the subcutaneous fat of adult Merlins. The egg results showed a shell thickness index of 1.21 compared to the 1.33 calculated for pre-pesticide eggs. Clutch sizes were as for pre-pesticide nests, but the number of nestlings per nest had fallen from 3.8 to 3.0. While DDE levels

Box 29
In his work on the effect of organochlorine contamination and eggshell thinning Derek Ratcliffe (see Box 27, *Chapter 10*) needed a measure of eggshell thickness without damaging the egg. Ratcliffe developed the equation (Ratcliffe 1967):

$$\text{Eggshell thickness index} = \frac{\text{Weight of shell (mg)}}{\text{Shell length (mm)} \times \text{Shell breadth (mm)}}$$

which is now known as the Ratcliffe Index.

For a comprehensive summary of Ratcliffe's work see, for instance, Ratcliffe 1970 or Ratcliffe 1993).

were lower than those in North American Peregrines, Temple concluded that Newfoundland Merlins 'now carry deleterious levels of DDE in their bodies, lay thinner shelled eggs, and produce young at a rate lower than in pre-pesticide times'. In considering the effect of the thinner shell, the comment of Rowan (1920-21 Part III) regarding the female Merlin's habit of tucking her feet under her eggs and so being likely to disturb one when leaving the nest, and of landing heavily among the eggs on her return (Fig. 20 in *Chapter 8*) has to be recalled. In addition to the direct effect of organochlorine contamination on eggshell thickness and therefore the possibility of shell breakage, Fyfe *et al.* (1976) noted that the contaminant burden of adult birds made them less aggressive against nest site intruders with a consequent increase in the risk of nest predation.

Use of the Ratcliffe Index is now widespread and has been subject to testing by other researchers. In interesting work by Fox *et al.* (1975), the research team used beta-particle backscatter to measure eggshell thickness. The Ratcliffe Index was shown to be highly correlated to the backscatter data, and the Index is now commonly used to assess shell thickness in the field. Perhaps more importantly, Fox (1979) presented data which allowed the level of DDE contamination in eggs to be assessed (within about 20%) by comparison with the Ratcliffe Index, thus allowing a measure of contamination to be assessed without the need for analysis, which would require expensive equipment. To do this Fox measured the Index for eggs from 110 nest attempts by *F.c. richardsonii* between 1969 and 1973 in Alberta/Saskatchewan. The nests were then grouped according to the Index and he then noted the number of nestlings from each group which reached the age at which they could be banded. The number of nests from which at least one nestling reached

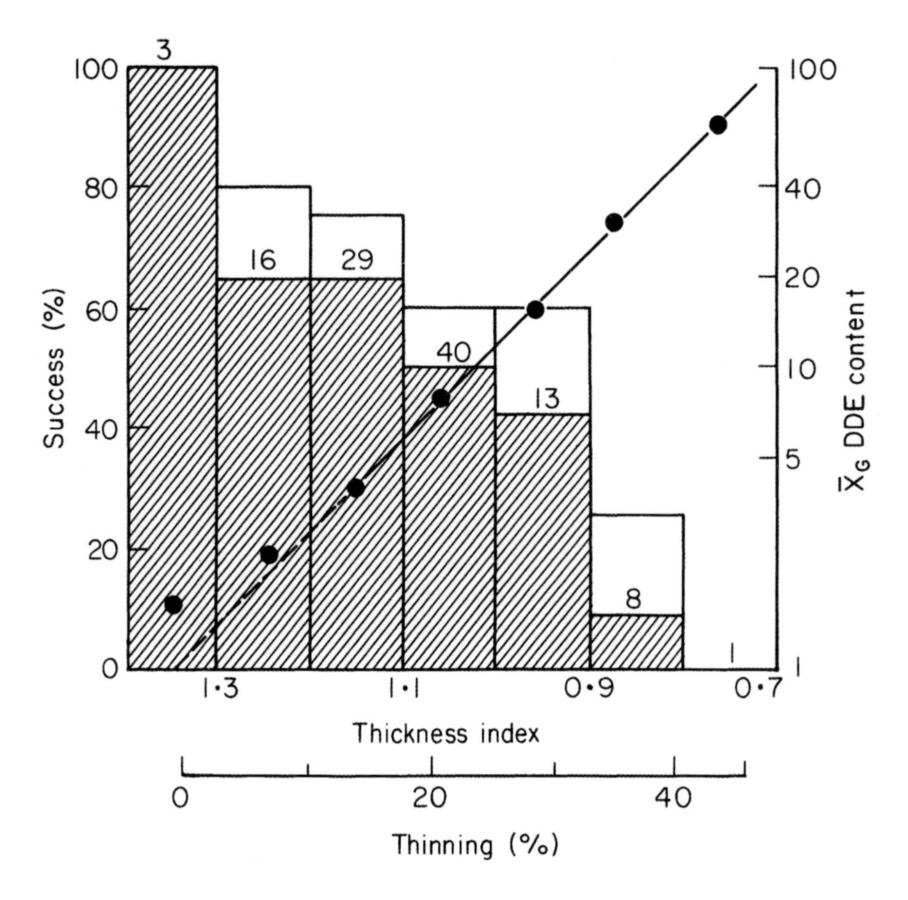

Figure 32
Variation of Ratcliffe Index with DDE content (ppm wet weight) and nesting success for 110 Richardson's Merlin nests. The shaded histogram is the percentage of nests in which all eggs produced nestlings reaching ringing age. The open histograms are nests in which at least one nestling reached ringing age. DDE content for a given Index class is represented by (●). Numbers above histograms are the numbers of nests in each Index group.
Reproduced from Fox (1979) © Wiley Online Library.

such an age, but which were not entirely successful was also noted. The DDE contamination was also measured for eggs of each Index. The result is shown in Fig. 32 which indicates the relationship between DDE level and the Ratcliffe Index, but also the relationship between the Index and Merlin breeding success.

In another study of the effect of organochlorine contamination, in this case on *F.c. richardsonii* on the prairies of south-eastern Alberta, Canada in 1975, Fox and Donald (1980) used the beta-particle backscatter technique to investigate five clutches. In this study the eggshells were even thinner (a mean of 1.00, range 0.88-1.31 for a group of the most contaminated nests), and the reproductive success lower, with 34 eggs resulting in 19 fledglings. However, Fox and Donald were able to establish that the behaviour of the female Merlin (chosen rather than the male as it is the female that determines the contamination of the eggs) showed no signs of abnormality.

Further confirmation that organochlorine contamination was responsible for the decline in breeding success, and therefore population, of the Merlin was provided by Becker and Sieg (1987b) who, in similar fashion to Fox (1971) studied museum eggs of *F.c. richardsonii* gathered in the northern prairies of the USA, and southern prairies of Canada. They noted the change in Ratcliffe Index from about 1.35 in the period 1865-1950 to *c.*1.10 in their own study of eggs from south-eastern Montana in 1980. Becker and Sieg found that, despite the cessation of use of Dieldrin and DDT in the area in the early 1970s, the contaminant levels in Merlins were still high enough to be influencing reproductive success. In a separate study, Noble and Elliot (1990) looked at organochlorine contamination of 27 species of Canadian raptors (using both egg and body tissue samples) and assessed both the critical levels of contamination required for adult death to occur (acute toxicity), as well as the levels that would result in reduced reproductive success. They found no evidence of acute toxicity – though they mention one recovered Merlin which had been ringed in Alberta and was found to have had a very high level of the organochlorine compound heptachlor epoxide (used as an insecticide) in its brain (see Henny *et al.* (1976) for full details of this bird). However, the results from the egg studies, which, of course, adversely impacte reproductive success, were alarming. From eggs gathered from the Canadian prairies in 1965-1972, 53% (N=165) had DDE levels above the critical limit (of 5mg/kg) – a single sample from Quebec was also above the limit; for eggs gathered in 1973-1979, 81% (N=159) were above the limit; and in 1980-1988, 36% (N=14) were above the limit.

Organochlorine contamination may affect reproductive success as a result of egg addling, or shell thinning resulting in egg breakage, but Risebrough *et al.* (1970) also suggest that an enhanced burden of DDE, and specifically PCBs, may cause delayed breeding. Delayed breeding has been shown to lead to reduced broods, and nestlings which fledge later are known to have increased mortality in their first winter. Delayed breeding is also of specific concern to northern breeding birds (particularly Arctic-breeding species) as the Arctic summer is short.

As well as the effect of organochlorine contamination (mainly accumulated in raptors as a result the widespread use of Dieldrin to control grasshopper populations, the insects being consumed by raptor avian prey species), Merlins in the North American prairie belt were also affected by a reduction in prey species because of changes in agriculture, with cultivation increasingly replacing grassland (Hodson 1976). The net effect of these pesticides and a reduction in hunting areas caused a serious

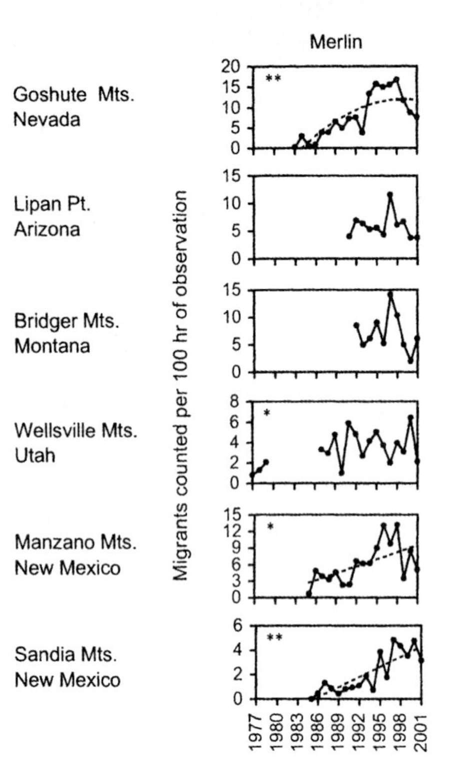

Figure 33
Migrant counts at look-out points in western North America. The increases noted at Goshute and Sandia Mountains are statistically significant at the 99% confidence level. Those at Wellsville and Manzano Mountains are significant at the 95% level. There was no statistically significant increase at Lipan Point and the Bridger Mountains.
Reproduced from Hoffman and Smith (2003).

decline in the Richardson's Merlin population. As noted above, Fox (1971) found that Merlins vanished from his own study area and this experience was mirrored across the prairie lands, to the extent that some authorities claim the species came close to extinction.

Given that Merlins accumulate contaminants from both their summer and winter prey (for the latter, see Schick *et al.* (1987) who measured organochlorine levels in Dunlin, which form a significant fraction of the Merlin winter diet, and other shorebirds in west Washington State) the outlook seemed gloomy. But despite initial fears, the North American population of Merlins began to increase. In a study based on spring and autumn counts at six well-known look-outs in north-eastern USA during the period 1972-1987 (though counts did not cover the entire period at all look-outs) Titus and Fuller (1990) showed a statistically significant increase in Merlin numbers. This result was confirmed by another study carried out by Hoffman and Smith (2003) who looked at Christmas Bird Counts and at migration counts at look-outs in the western USA from New Mexico and Arizona, through Nevada and Utah to Montana (Fig. 33). While it is likely that the increases seen by Hoffman and Smith relate to nominate and Black Merlin populations, an increasing population of Richardson's Merlin has also been seen in Canada (Houston and Hodson 1997, Downes 2003).

The recovery of *F.c. richardsonii* has been ascribed to a combination of factors. One is the increase in reproductive success following the cessation of use of organochlorines. Another has been an increase in the species' winter range (James *et al.* 1987a) which, as already noted (see *Chapter 9*: *F.c. richardsonii*), is believed to have been associated with an increase in the

Female Taiga Merlin.
Tom Grey.

population of Bohemian Waxwings. Merlins also increased their breeding range in the decade from 1995-2005, occupying virtually all of the state of Maine, moving into northern areas of the states of New York and New England, and into Vermont and New Hampshire. In many instances the expansion was driven, in large part, by a move to urban areas, something which has yet to be seen in Palearctic Merlin populations. The range increases coincided with population increases. Kirk and Hyslop (1998), in a study of Canadian raptor populations, noted that for Canada as a whole the population had increased (by about 1%) over the period 1966-1984, but that this increase was mainly due to a larger increase in the years 1985-1994, i.e. after cessation of the use of organochlorine pesticides. The data, obtained from a combination of Christmas Bird Counts, migration counts and specific research programmes, indicated increases for both nominate and *F.c richardsonii*, though the increase for the nominate population was more pronounced for the period 1985-1994. A later study (Sauer *et al.* 2012), based on Breeding Bird Surveys, but excluding Alaska and Canada north of about 60°, indicated an average increase of 5.5% in the overall North American population (4.8% in Canada, 11.5% in the USA) in the decade 2001-2011 (Sauer *et al.* 2012) with many individual Province or

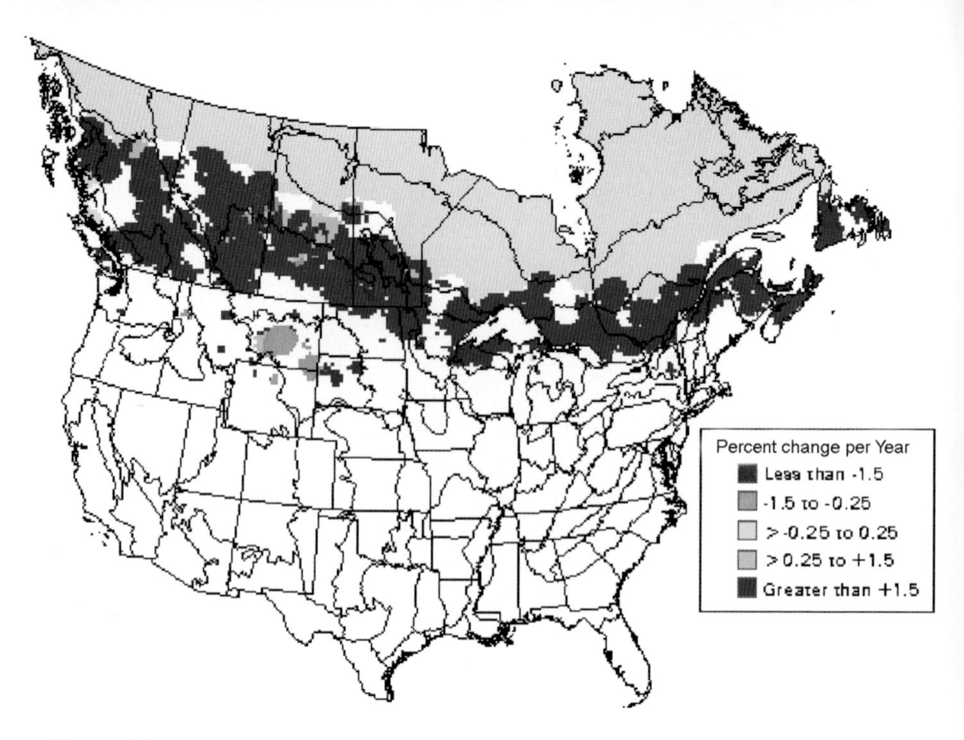

Figure 34
Trends in the increase of Merlin population across North America from 1996-2011.
Reproduced from Sauer *et al.* (2012).

State populations stable or increasing. Though most obvious because of the movement into urban areas, population increases have been seen for both the nominate bird and *F.c. richardsonii*: the population of *F.c. suckleyi* is believed to be stable (Fig. 34).

The overall North American population was estimated (Rich *et al.* 2004) to be 650,000 birds, based on the assumption that 50% of a world population of an assumed 1.3 million birds was to be found in North America, with half the population in the USA and half in Canada. (As a digression, Millsap and Allen (2006) use an estimated US population of 325,000 to recommend an annual wild harvest (the taking of birds for falconry) of Merlins of 1%, but note that the actual harvest in 2003/2004 averaged 50 birds/year. The present author considers the suggested Nearctic and world populations are significant overestimates of the true populations.

Iceland and the Faeroes

Nielsen (1986) calculated the Ratcliffe Index of four clutches of Merlin eggs collected between 1900 and 1940 and held in the Icelandic Museum of Natural History. The mean eggshell thickness was 1.27 (range 1.21-1.31). By comparison, 15 clutches collected between 1952 and 1985 had a mean Index of 1.10 (range 0.97-1.26) indicating a 13% reduction since organochlorine pesticides were first used on the island. Nielsen raised a concern with respect to the Icelandic Merlin population, noting that the falcons were, in general, migrating to Britain during the time of his study and that the British Merlin population was continuing to decline despite the banning of DDT and related pesticides. Nielsen also noted that many of the prey species which Icelandic Merlins consumed during the breeding season also migrated to Britain for winter.

Assuming the Merlin breeding density of 2.1 pairs/100km^2 derived from Nielsen's study, and assuming also that the coastal land area of Iceland which can support breeding Merlins constitutes about a third of the total land area of approximately 102,000km^2 then the population at the time of Nielsen's study would be about 700 pairs, consistent with Nielsen's view that it was 500-1000 pairs. Nielsen considered that at the time of his study the population was stable (but with concerns for the future as noted above). Since his study, the population in Britain has increased and there is no reason to suppose that improved reproductive success has not occurred in Iceland. In their review of raptor populations in Europe, North Africa and the Middle East, Mebs and Schmidt (2006) confirmed the Nielsen population estimate of 500-1000 pairs (though they quote a value from Génsbøl and Thiede 2004). Nielsen also considers (Ó.K. Nielsen, pers. comm. 2014) that it is likely the Merlin population has recovered in much the same way as that in Britain (see below) which would mean that the upper figure of the Mebs and Schmidt (2006) estimate would be the more likely present population.

The Faeroese population is 25-50 pairs, probably at the lower end of this range (S. Olafson and J-K. Jensen pers. comm. 2014).

British Isles

In a report on the status of breeding birds in Great Britain and Ireland, Parslow (1967) noted that over the previous 30-60 years there had been a general decline in the Merlin population across the British Isles, largely as a result of persecution, though no estimate was made. Persecution was carried by game keepers on moorland managed for the shooting of grouse, and was indiscriminate (e.g. McMillan 2011). Merlins were killed despite their minimal impact on the grouse population due to the fact that they only take the occasional grouse chick.

Parslow also noted that in recent years the decline had been more marked. While, as noted above, the main emphasis on the near catastrophic decline in raptor population as a consequence of organochlorine contamination concentrated on the Peregrine Falcon, the effect was an equally important factor in the decline in the breeding success of the Merlin: it was contamination which had precipitated the decline referred to by Parslow.

That contamination was the major cause can be seen in the data of Newton *et al.* (1978) which compared clutch size and fledgling success in the period 1961-1976, a period which saw a decline in the usage of organochlorines (Table 25). While the mean clutch size did not vary over the 15 year period, the increase in the number of nests from which at least one nestling successfully fledged, and mean number of nestlings successfully fledged was statistically significant, at the 99% and 95% levels respectively.

Period	Mean Clutch Size	Percentage of Nests in which at least one nestling fledged	Mean number of Fledglings
1961-1970	4.44	50	3.00
1971-1973	4.33	60	3.06
1974-1976	4.42	75	3.57

Table 25 Mean clutch size and number of fledglings in Merlin nests in northern England during the period 1961-1976. From Newton *et al.* (1978).

While the data of Table 25 reflected the loss of reproductive success in northern England, the losses were seen across the British Isles. In the Orkney Islands, off Scotland's northern coast, Meek (1988) found that the mean reproductive success, defined as the proportion of Merlin pairs successfully raising one chick, fell from 48% (1975-81) to 29% (1982-1986). Meek also found that the mean brood size in successful nests fell from 3.3 during 1975-80 to 2.5 during 1981-87.

However, while the reproductive success of the falcons had increased following the banning of the use of DDT and related pesticides, the decline of the Merlin had continued. At one site in Wales, where historically eight Merlin pairs had bred in the mid-70s, only one remained in a study by Roberts and Green (1983) in 1982. They noted that in this case no single cause for the population reduction could be identified, the decline seemingly a combination of predation (largely by humans), fire, poor weather and organochlorine contamination. Roberts and Jones (1999) considered that a fire which had destroyed several preferred nest sites was responsible for a decline in their study area in north-east Wales.

In another study (Newton *et al.* 1981) noted that recent studies across Britain had suggested a decline, and that the Merlin had almost disappeared from the Peak District National Park where the general habitat was, in principle, ideal for the species. Newton and co-workers considered several potential causes of decline. One was the loss of heather cover in favour of grassland for sheep rearing. Heather is essential for ground nesting, and traditionally Merlins in the Park had been ground nesters in the absence of trees (and, therefore, stick nests). However, counting prey species on heather and grassland suggested little change in prey density, and although it had been possible to identify lost nest sites in heather areas which had been used previously, but were now lost to grassland, it was considered that overall the loss of heather had been marginal, and seemed not to be the cause of the population decline. Recreational use of the Park was also considered. Since the 1960s there had been a significant rise in the numbers using the Park for walking and other outdoor pursuits, numbers visiting the National Park Information Centre rising by almost 400% between 1965 and 1980. Disturbance by humans was therefore a possibility, but it seemed footpath use predominated, most visitors foregoing the 'pleasure' of walking through dense heather in favour of the less arduous grassy tracks. Overall Newton *et al.* (1981) were drawn to the conclusion that the effects of organochlorine contamination were largely responsible for the continuing Merlin population decline. That Merlins were contaminated by organochlorine was confirmed in a later study (Newton *et al.* 1982).

Newton and Haas (1988), in a study of eggs gathered between 1964 and 1986, provided further evidence that the Merlin was at risk from continuing contamination, with DDE, HEOD (from Dieldrin and Aldrin), PCBs and mercury all present. While levels of the organochlorine pollutants had fallen during the study period the Merlin remained 'the most heavily contaminated of British raptors'. Newton and Haas noted that the shell thinning Index was negatively correlated to DDE levels (confirming findings from other studies both in Britain, e.g. Newton *et al.* 1982, and in North America, e.g. Hodson 1976), while the overall burden of organochlorine contamination showed the same variation, i.e. DDE was the dominant contaminant with respect to shell thinning. Since DDE is also the most persistent of the organochlorines in the environment this was a worrying finding. However, there was no negative correlation between organochlorine contamination and brood size, which is curious given the evidence of enhanced shell thinning: Newton and Haas (1988) conclude, that at present levels, organochlorine contamination is no longer the main reason for brood reduction, the continuing effect being outweighed by other factors. What was clear, however, was that there was a negative correlation between brood size and mercury levels. Equally interesting was the fact that Merlins in areas which had not experienced significant use of organochlorine insecticides (e.g. the western Highlands of Scotland) were as contaminated as those from areas which had: the conclusion is that Merlins can pick up contamination from prey captured both at wintering sites, and from summer prey which has wintered in contaminated areas. A later study (Shore *et al.* 2002) confirmed that Merlin were the most contaminated of British raptors, but noted the continuing decline in organochlorine levels in Merlin eggs.

Meek (1988) also looked at the potential causes of Merlin population decline in the Orkneys where the Merlin, once described as 'very common' (though this was, of course, a relative term), had reduced in numbers such that only 5 pairs nested in 1986 and none produced fledglings. In Britain the covering of large areas of upland with conifer plantations in the 1960s and later years, continuing a trend which had begun half a century before, had potentially reduced the available habitat for the falcons. The Orkneys offered an opportunity to see if that could be an explanation as there had been no commercial forestry on the islands. Theywere also free of mammalian predators (apart from feral cats, domestic dogs and Eurasian Otters, though the latter are concentrated at the coast) which had been suggested as another possible cause for Merlin decline. Meek found that 48% of Merlin nests failed at the incubation stage, only 5% failing after hatching. As egg failure looked the likely cause of decline in reproductive success, Meek analysed sample eggs. The results showed organochlorine burdens, but at lower levels than were still occurring on the British mainland, but mercury levels which were higher than those seen across most of the mainland (confirming the result of Newton and Haas (1988) who had also found much higher mercury levels in Orkney eggs). The Ratcliffe Index was also low at 1.08 (mean for ten eggs) in the period 1982-1986. Meek considered other potential causes for the population decline: there had been changes in land usage, with the ploughing of moorland to produce cattle pasture, but only one historical Merlin nest site had been lost; heather burning, for Red Grouse management and sheep rearing, had similarly altered land usage, but again with no apparent impairment of available Merlin nest sites; human disturbance seemed minimal and although there were known losses to feral cats, these, too, seemed minimal in comparison to predation of nests by Hooded Crows which, though the crow population had increased, did not seem to account for nest losses; and although the study period included a succession of cold springs, the weather overall was not unduly harsh. It therefore seemed most likely that contaminants, particularly by mercury, were the cause of egg failures.

The diet of Orkney Merlins shows a much higher fraction of House Sparrows (17% of total biomass) and Meek noted that the barley traditionally grown on the Orkneys was treated with organo-mercury. Ellis and Okill (1990) also found very high levels of mercury in Merlins on the Shetland Isles, which lie to the north of the Orkneys (again confirming the result of Newton and Haas (1988)). The Shetland Merlin population there had declined by 50% in the years 1981-1983, with a breeding success of only 50% in 1982 (compared to an average of around 78% in the years 1976-1980). 97% of failed nests failed at the egg stage, with egg breakage apparently a major cause, shell fragments suggesting many breakages by the adult birds: in addition to this some eggs were addled. The mean Ratcliffe Index during the study period (1981-1985) was 1.14. Ellis and Okill considered adverse weather and predation as being unlikely causes of the decline, and note that the levels of chemical contaminants other than mercury were lower in the falcons than in their British mainland cousins.

The Orkney and Shetland mercury results were curious. Newton and Haas (1988) had found a negative correlation between mercury level in eggs and brood size for eggs gathered in mainland Britain, but the Orkney and Shetland results did not fit the regression: Merlin pairs on the islands have a range of brood sizes which covered the entire spectrum of mainland pairs despite having mercury levels significantly higher. Newton and Haas could not explain this difference, though they did suggest it could have resulted from the specific mercury compound used in the two areas. In a later study (Newton *et al.* 1999) noted the decline in organochlorine levels in Merlin eggs collected from across Britain, but that mercury levels, which had declined up until the mid-1980s, had then increased before showing a further fall towards the mid-1990s. Looking at individual regions, the levels of mercury in eggs had remained essentially static in the Orkneys and Shetlands (pre-1987 and post-1986) while those of several other regions – Wales, north-east England and north Scotland – had shown significant rises, the eggs from there having levels above those of the northern islands. The odd fluctuations in Merlin egg mercury levels will need continuing research to decide if they are a cause for concern.

In their study of Merlin population decline, Newton *et al.* (1978) had also noted that while the loss of potential nest sites reduces the number available, there were no data on the number of sites which were not historically suitable, but had become so. However, the species' requirement for a precisely acceptable habitat meant creation of acceptable Merlin nest sites by chance was unlikely, with a further implication that in the absence of conservation measures Merlin numbers might continue to decline. But despite this concern and the evidence of continuing high levels of egg contamination, the number of Merlins in Britain began to rise.

Bibby and Nattrass (1986) estimated the Merlin population of Great Britain by asking field workers to count birds in their areas. Coverage was not complete (it was claimed to be complete in the Orkneys, Shetlands and western Isles, in Wales and northern England, but patchy across much of Scotland and absent south of the Peak District National Park), but apart from mainland Scotland it did cover those areas which historically had been strongholds of the species. The result was that the population in Wales was 40-45 pairs, with 180 pairs in northern England, and 100 pairs in the Scottish islands. On mainland Scotland 133 pairs were reported, but Bibby and Nattrass reasonably claim that this probably underestimated the total by as many as 200-300 pairs because of the lack of coverage. Their estimate was, therefore 550-650 pairs in 1983/84, a figure that seemed to confirm a continuing decline.

But by the 1990s the population had begun to recover. Parr (1991) noted that the Merlin population of an area of central Wales had increased, from perhaps as few as 5-10 pairs in 1970-1975 to 10-20 pairs in 1986-1989, and that the reason was, in part, the use of stick nests in trees at the edges of forestry plantations. In this area of Wales the Merlins has historically utilised stick nests in moorland trees, but the new plantations were offering new opportunities. Parr considered that if the results were to be replicated across Wales then the population might have increased from 40-45 pairs (Bibby and Nattrass 1986) to perhaps 60-70 pairs. Parr (1994) confirmed this estimate in a study of breeding areas across Wales, though Williams and Parr (1995) considered the population was 80-90 pairs on the basis of better survey coverage.

Little and Davison (1992) and Little *et al.* (1995) showed a similar change of nest site in a study in the Kielder Forest in northern England where the Merlin population had increased from 10 pairs in 1982 to 29 pairs in 1991 as a result of the use of stick nests in forest edge trees. (Kielder Forest is England's largest manmade woodland and chiefly comprises Sitka Spruce.) Data in Little *et al.* (1995) show (Fig. 35) that use of traditional nest sites (ground nests among heather, or nests on rock outcrops) had remained essentially static, the population increase being driven by stick nest utilisation in forest edge trees. The stick nests were those of Carrion Crows which began utilising the Sitka Spruce trees once they had grown to 4-5m in height. Interestingly, the use of forest edge trees was associated with a shift in hunting habitat, the Merlins utilising not only the traditional heather moorland, but grass moorland close to the forest edge and, therefore, the new nesting sites (Fig. 36). By studying the nests of Merlins which used the different hunting habitats, Little and co-workers were able to show that there was no difference in the reproductive success of the pairs utilising either moorland habitat.

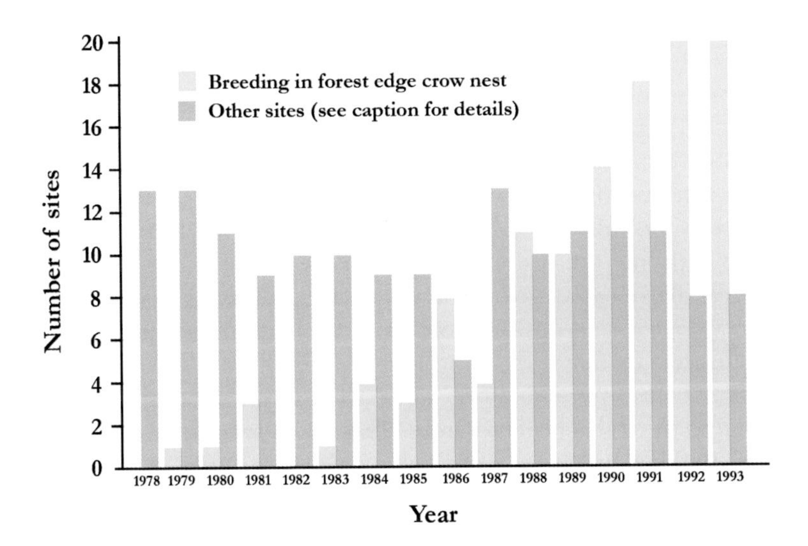

Figure 35
Merlin nest sites in the Kielder Forest, Northumberland, England during the period 1978-1993. FCN are stick nests in trees at the forest edge. OTN are other nest sites.
Redrawn from Little *et al.* (1995).

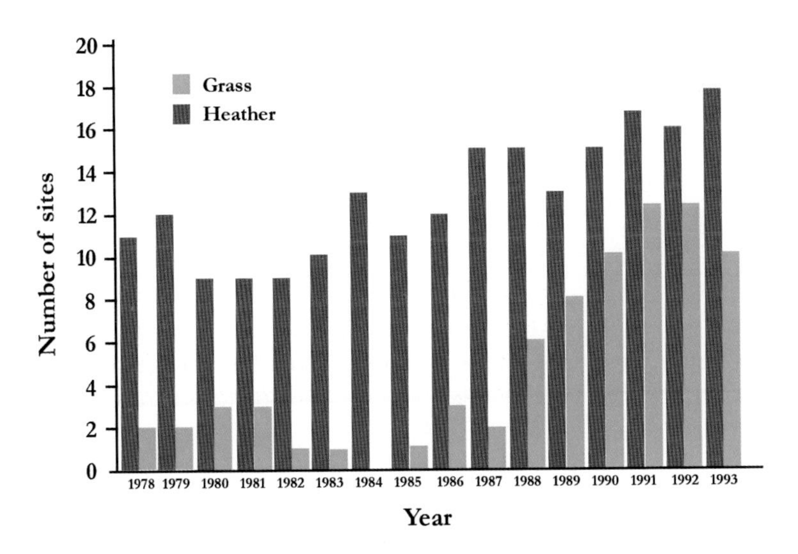

Figure 36
Merlin hunting habitat associated with nests sites in the Kielder Forest, Northumberland, England during the period 1978-1993. Habitats are heather and grass moorland within 4km of each nest site.
Redrawn from Little *et al.* (1995).

However, in a separate study in the Kielder Forest, Petty *et al.* (1995) found that Merlins did not respond to eruptions of Common Crossbills and Siskins arising as a consequence of peak conifer seed production in the forest, keeping to their 'normal' diet of moorland birds. A consequence of not taking advantage of this increased abundance of potential prey meant that the population of Merlins remained stable, whereas the population of the Eurasian Sparrowhawk cycled in synchrony with the prey resource. A potential consequence of this finding is that while increased afforestation may offer Merlins the opportunity to utilise new nesting sites at the forest edge, the falcon's hunting techniques are not adapted to forest dwelling and so they may not be able to take advantage of the potential prey species which breed in the forest, preferring traditional moorland areas. This suggestion is in keeping with the finding of Newton *et al.* (1984), shown here in Table 26.

	Sheep Grazing	**Heather Moor[1]**	**Mixed open land/forest**	**Forest Plantation**
	N (%)	**N (%)**	**N (%)**	**N (%)**
Open Country species	85 (88.5)	889 (88.0)	371 (80.8)	369 (84.6)
Woodland species[2]	11 (11.5)	121 (12.0)	88 (19.2)	67 (15.4)

Table 26 Number and percentages of open country and woodland species prey items in the diet of Merlins according to the predominant habitat found within 1km of the nest during the period April to July. From Newton *et al.* (1984).

Notes:
1. The heather moorland was managed for Red Grouse.
2. Woodland species included those associated with scrub and woodland edges.
3. The difference in the two classes of prey species is statistically significant at the 99% level for all four habitats.

What is clear from Table 26 is that the observed increase in Merlins choosing stick nests at the edge of new forestry plantations has not been accompanied by a marked change in the falcon's diet. A continued spread of conifer plantations, which are usually planted on moorland rather than land which is more economically useful, may, therefore, result in an increase in nesting opportunities, but may also limit the availability of habitat for the falcon.

Rebecca and Bainbridge (1998) revised the population figure for Britain (i.e. England, Scotland and Wales), again using observers to count birds in areas expected to hold breeding Merlin pairs. The estimate was 1100-1500 pairs (at the 95% confidence level), about double the 1983/84 estimate and a very encouraging outcome. Finally, Ewing *et al.* (2011) used the same surveying techniques as Rebecca and Bainbridge (1998) to revise the population figures up to 2008. The new estimates suggested 733 breeding pairs in Scotland (see also Etheridge 2013), 301 pairs in England, 94 pairs in Wales and 34 pairs in Northern Ireland making a total of 1162 pairs in the United Kingdom, but with a range of 891-1462 (at the 95% confidence level). However, the significant figures were that the population had reduced by 13% overall since the 1998 survey, with a 7% reduction in Scotland and a 25% decrease in England, but a 16% rise in Wales and no change in Northern Ireland. Ewing and co-workers discuss potential causes of the reduction in Merlin numbers. Petty *et al.* (2003) note that in the Kielder Forest in northern England Goshawks are considered a significant cause of decline in Eurasian Kestrel numbers as they predate the small falcon, but, as noted above (see *Chapter 11: Foes*) the fraction of Merlins in the Goshawk diet is much lower and so may not be a proximate cause of Merlin population decline. However, that view has since been challenged as the Merlin population of the Kielder Forest has disappeared, and the reason is considered to have been the long-term effect of Goshawk predation (A. Heavisides, pers. comm.).

Ewing *et al.* (2011) state that the populations of the primary Merlin diet, both in summer and winter, have declined in Britain, with potential influences on breeding success and winter survival. They also consider that heather burning for the management of moorland for grouse shooting may be having an impact, as may climate change, as the Merlin is at the southern end of its breeding range in Britain and so may be more susceptible to the effects of change. While the population decline is not as catastrophic as it was as a result of the organochlorine contamination, changes in land usage in an increasingly crowded Britain have the potential to cause further declines: as has been shown in North America, cultivation may cause a reduction in prey species, the Merlin being unable to breed in farmed areas which do not offer a habitat for their primary prey.

In his study of Merlins on the Wicklow Mountains, McElheron (2005) noted that the less intensive agricultural practices in Eire, in comparison to those of neighbouring Britain, meant that fewer pesticides were used, but that the Merlin population still declined. A larger threat was that posed by the increase in subsidies for sheep farmers in the 1960s and 1970s which resulted in the indiscriminate burning of large areas of heather moorland (McElheron refers to the 'mindless destruction' by a 'small number of farmers') and a consequent reduction in the habitat of both the Meadow Pipit and the Merlin, depriving the latter of both prey and nesting sites: the few Merlin which remained, turned to woodland species for prey and stick nests in isolated tree clumps or at the forest edge for breeding. The concerns expressed by McElheron were echoed by data from a study, in which he collaborated (Norriss *et al.* 2010), which examined the ecology of several Merlin breeding areas across Ireland, including the Wicklow Mountains. This showed that the productivity of the Wicklow Merlins was 1.85, i.e. that only an average of 1.85 fledglings were raised per breeding pair. The range of productivity across Ireland was 1.27-2.84, the lower figure from nests in County Kildare, the higher from those in an area of County Donegal. The mean across Ireland was 2.23.

Neither McElheron (2005) nor Norriss *et al.* (2011) quote a figure for the Irish Merlin population. Greenwood *et al.* (2003) suggested a total of 200 pairs in Eire, while Mee (2012), collected data for all Irish raptors and considered it could be as high as 250 pairs (Eire + Northern Ireland) based on data collated in 2010. However, Lusby *et al.* (2011) noted that the winter of 2009/2010 was the most severe for 50 years and was likely to have exacerbated a decline in passerine numbers, particularly of upland species (for instance Meadow Pipits) which form a significant fraction of the Merlin diet. Fernández-Bellon and Lusby (2011b) studied the breeding success of Merlins in 2010 and 2011 which followed the harsh winter and found that at 14 nest sites only 0.7-1.3 chicks fledged, a breeding success rate which is below the 2.6 fledglings/breeding attempt which Bibby (1986) considered was required to maintain a stable population (though Rebecca *et al.* 1992 noted that the Scottish Merlin population seemed stable despite the breeding success rate being only 2.2 fledglings/breeding attempt (**Box 30**). The breeding success measured by Fernández-Bellon and Lusby (2011b) therefore raises concerns for the future of the Merlin population, though the present best-estimate is 200-400 pairs in Eire and a further 32 pairs in Northern Ireland (S. Newton, pers. comm.).

The wide estimate for Eire is due, in part, to the difficulty of assessing the population, as Lusby *et al.* (2011) note. In a novel experiment (Fernández-Bellon and Lusby 2011c) researchers tried to use the playback of conspecific calls to elicit a response from Merlin pairs and hence improve surveying results: in practice only 28% of pairs responded

Box 30
Mebs (1971) suggested the following equation to calculate the required productivity of Peregrines to maintain the population level:

$$f = \frac{2m}{(1-q)(1-m)}$$

where

q is the mortality (1.0-survival) of birds dying before reaching breeding age,
m is the mortality (1.0-survival) of birds during their breeding life (assumed to be 2-6 years)
and f is the required productivity for population stability.

Assuming the equation holds for any species which breeds at two years of age and has a breeding life span of about four years, the equation can be used to define the required productivity for Merlins. But even if that assumption is made (and it seems reasonable), choosing values for q and m makes an apparently simple job very complicated. Heavisides (1987) suggests q=0.605, while Fig. 28 suggests that m=0.25 (certainly for males, and approximately for both sexes). Using these figures gives a value of f of 1.69. If q is higher (i.e. the survival of first winter Merlins is lower), then f is also higher (q=0.7, m=0.25, f=2.22): if m is lower (i.e. the survival of older Merlins is higher), then f is lower (q=0.605, m=0.2, f=1.27). The problem, so apparently simple to solve, is in choosing correct values for q and m. or Ratcliffe 1993).

vocally, most responses being discrete and silent, so a negative response was not evidence of absence and the use of playback failed.

Overall it seems the Merlin population of the British Isles is stable or increasing slowly, though with regional variations that mean the population is decreasing in places, but with such a limited population the species remains vulnerable. Rebecca (2006) noted the decline in Merlin population in his study area as a consequence of commercial forestry, and the most recent British Wetland Bird Survey (Austin *et al.* 2014) noted a significant fall in populations of some waders, including Dunlin and Redshank (the former showing a record low, the latter the lowest number for 30 years) each of which is an important prey species for Merlins overwintering at estuaries. If such reduced numbers are sustained, or further reduced, there may be implications for winter survival of the falcons and, consequently, on the population, which would be exacerbated if reductions in breeding range, as a consequence of land usage changes, were to increase.

Scandinavia

The problem of organochlorine contamination affected the Merlin population of the Scandinavian countries just as it had in North America and Britain, but again the population appears to have recovered, though the difficulties of estimating the population in the wildernesses of northern Fennoscandia make any attempt subject to sizeable errors.

On a local level significant work has been carried out by Christopher Wiklund in Sweden, some of which has already been mentioned in the preceding chapters. One factor which Wiklund (1995) noted as affecting the population of Merlins in any given area is the effect of predation on the LRS of females which, of course, affects the potential increase in Merlin numbers. Wiklund, in one of a series of papers which used data from studies over several years in two Arctic National Parks, found that while breeding life span was the major contributor (62.3%) to the LRS of female Merlins, the contribution of predation was 31.1% (other factors being mean clutch size, mean brood size and the influence of the egg laying period). Given that Merlins are also dependent on the availability of nests for breeding, and that the primary providers of nests are Hooded Crows, a Merlin nest predator, it seems the ecology of the Merlin, as with other species, is a delicate balance which can be easily tipped by causal interference.

In his study of Merlins in the two National Parks, Wiklund (2001) found a negative correlation between the return of male Merlins and the number of pairs which had attempted to breed in the previous year, with the population of rodents being the apparent mechanism. A similar effect, but one which was much less statistically significant, was seen in the female Merlin population. The rodent population of that area of Sweden was cyclic with a four-year period (although the cyclicity failed during the latter period of Wiklund's study for unknown reasons). The study indicated that if many pairs of Merlin attempted to breed in a given spring, males were less likely to return the following year, one possible reason being that they were aware of the rodent population cycle and were also aware that competition for territories would be high. For females the same reasons might apply, but if breeding was successful the previous year the imperative to return might be an over-riding factor.

Wiklund does not quote a population figure for Merlins in Sweden.

The current best estimates for Fennoscandian Merlins are:

Mebs and Schmidt (2006) quote a figure of 2500-6500 breeding pairs for Norway based on a Bird Life International estimate in 2004, which accords with the data of Gjershaug *et al.* (1994) who suggested 2000-6000 pairs at the time of publication, with an assumed slight increase since then. The Mebs and Schmidt (2006) figures were considered correct by Heggøy and Øien (2014), who also state that 30-100 Merlins remain in Norway during the winter, mostly in the Lista and Jæren areas in the far south of the country.

Svensson *et al.* (1999) consider the Merlin population of Sweden to be 4200-5700 breeding pairs.

In a study by Järvinen and Koskimes (1990) on Finnish raptor populations, the Merlin population was noted to have declined such that the species had been reclassified from 'out-of-danger' to 'in need of monitoring', the decline probably precipitated by organochlorine contamination, though deterioration of habitat in wintering quarters was also considered to have had an effect. However, the population has now grown to 2400-3600 breeding pairs (pers. comm. J Honkala, Finnish Museum of Natural History, 2014).

Zimin *et al.* (2005) consider the best estimate for Karelia (Russian Fennoscandia), including the Kola Peninsula and the White Sea coast to Arkhangelsk, to be 700 pairs, and to be increasing slowly. Zimin and co-workers believe the majority of the Merlin population is in the northern Kola region: Koryakin (2005) considered there were 100-200 pairs in the Kola area close to Murmansk where the Merlin is the most abundant falcon. In southern Karelia Zimin *et al.* (2005) consider the Eurasian Kestrel to be more abundant.

Baltic States and European Russia

Mebs and Schmidt (2006) quote values of 5-10 breeding pairs for Lithuania, 30-100 pairs for Latvia, and 20,000-30,000 pairs for European Russia. Elts *et al.* (2013) suggest 30-60 pairs for Estonia, which represents a tripling of the population since the previous estimate in 2002 (Elts *et al.* 2003) and they estimate that between 5 and 30 Merlins remain in Estonia over winter.

Russia

Forsman (1999) estimated the Merlin population of the Western Palearctic at 35,000-55,000 of which 75% was in Russia, i.e. 25,000-40,000 pairs. Morozov *et al.* (2013) note that the latest estimate is 20,000-30,000, quoting Bird Life International (2004) as well as Russian sources. Estimating the population for Asian Russia is much more difficult as the

area is vast, as is the range of the Merlins, but the breeding density of the falcons is not constant across that range.

Morozov *et al.* (2013) quote breeding densities for some *oblasts* (administrative districts) of Russia, but the majority of these lie within European Russia. Only the Yamalo-Nenetsk Oblast lies within Siberia, and for that the population quoted is 11,200 individual falcons, which probably converts to about 5000 breeding pairs given the likely age structure of the population. For other areas, Morozov *et al.* (2013) quote breeding density for some relatively small areas, but much of the information is based on the number of Merlins or Merlin pairs which were seen per kilometre of exploration which is difficult to convert into a useful pair density.

For specific areas of tundra, Morozov and co-workers quote values which range from 9-12 pairs/100km^2 on the Bol'shezemelskaya tundra, but for other tundra areas the figure can be ten times lower. In the taiga, the density will be even lower, perhaps 0.6 pairs/100km^2 for the northern taiga, and the same or a moderately higher number for that in the south, but only 0.1 pairs/100km^2 in the middle taiga. Mountain areas provide a more difficult problem, as in the high mountain areas there may be no Merlins, while in biodiverse river valleys in the foothills the breeding density can be much higher. Morozov *et al.* quote values of 1-3 pairs/100km^2 for rivers of the Polar Urals and for the Putorana Plateau. Within the steppe zone the breeding density appears higher, in some places much higher, with values of 20-50/100km^2 being quoted for some areas. But again the problem is one of deciding how representative an area is of the steppe as a whole.

The only available figures covering large areas which can be viewed as representative, or nearly so, for Siberia as a whole are that of Morozov *et al.* (2013) for the Yamalo-Nenetski Oblast which suggests a population of about 5000 breeding pairs, and that of Vartapetov (1998) who quotes a figure of 40,000 falcons (not pairs) for the northern taiga of the West-Siberian Plain, an area which encompasses part of European Russia as it extends from the Baltic states/Finland to about 80°E. Vartapetov puts a very large range on his figure – 20,000-80,000 – which suggests a population of the non-European area of the plain of 0 (!) to 40,000 birds, or no more than 20,000 pairs.

To construct an estimate, and limits on that estimate, the following exercises have been undertaken. First and simplest, is to assume a breeding density for Siberia as a whole, allowing for low figures in the

Opposite
Female Central Asian Merlin.
Igor Karyakin.

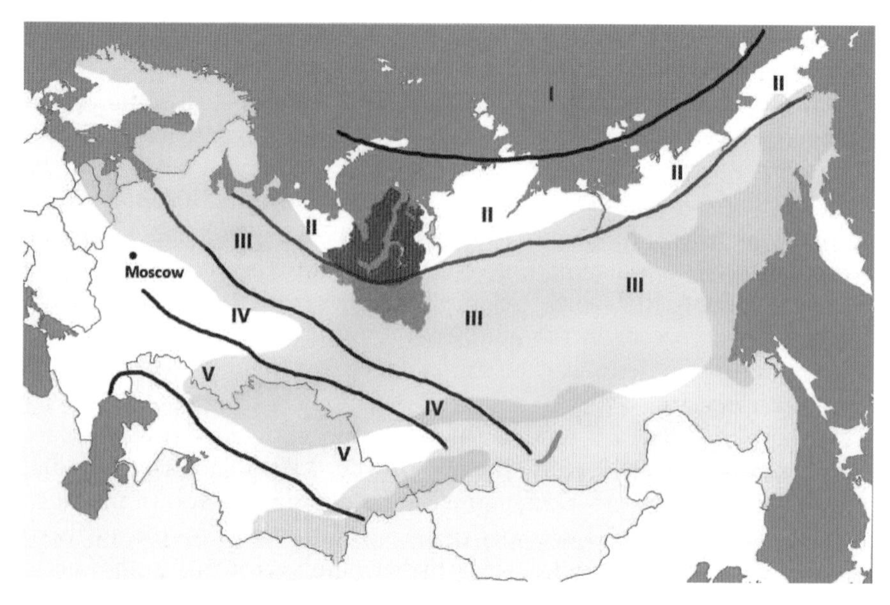

Figure 37
Biome/Biotic structure of Siberia.

I	Polar Desert	Red	Yamalo-Nenetski Oblast
II	Tundra/Forest Tundra	Brown	Mountains
III	Taiga	Green	Merlin Range
IV	Forest-Steppe		
V	Steppe		

taiga, with higher ones to the north and south. From the data of Morozov *et al.* (2013) the range of breeding densities would seem to be 0.5-1.0 pairs/100km^2 which would yield a total population of 130,000-250,000 falcons.

The second estimate attempts to break down the area into smaller regions with the aim of constructing a more valid picture of the entire area. Fig. 37 is an approximate breakdown of the biomes/biotic structures of Siberia with the Yamalo-Nenetsk Oblast marked. If it is assumed that this oblast is representative of northern Siberia as a whole, then the breeding population of northern Siberia is about 25,000 pairs. If that population is correct then the higher figure of Vartapetov (1998) of about 20,000 breeding pairs in the middle taiga of the Western Siberian Plain looks too high, with 5,000 pairs looking more reasonable, particularly if it is assumed to cover both the middle and southern taiga of the Plain. Assuming that figure on an area basis for the remaining taiga leads to a total taiga population of about 35,000 pairs. Assuming the steppe as a whole has a density of 2 pairs/100km^2 which seems reasonable against the data of Morozov *et al.* (2013) then the Merlin population of 1,500,000km^2 of steppe is 30,000. Calculations for the mountain zones are much more difficult, but assuming 1 pair/100km^2 as a starting point leads to a population of 25,000 pairs. The total Merlin population of Asian Russia is then 115,000 pairs. The error on this figure is considerable, with probably at least a 50% chance of overestimating the population. Based on this assumption of potential error rate the population figure would translate into 50,000-120,000 pairs, and a total population of 150,000-300,000 falcons. It is much less likely that the figure has been underestimated. Note that the agreement between the figures derived from the two estimates is hardly chance as the choice of breeding densities in the second assessment is never very far removed from that assumed in the first, and is therefore no indication of accuracy.

But assuming the figure for Asian Russia is of the right order, the total Palearctic Merlin population would then be 180,000-350,000. Assuming that this actually does represent half the world population then the total number of Merlins is about 500,000 birds, with a probable range represented by ±150,000.

The next generation of Merlins.

References

Note: Titles in parentheses have been translated from the original language.

Abuladze, A., *Materials towards a Fauna of Georgia – Issue VI: Birds of Prey of Georgia*, Ilia State University, Tbilisi, 2013.

Albuquerque, J.L.B., Observations on the use of rangle by the Peregrine Falcon (*Falco peregrinus tundrius*) wintering in southern Brasil [sic], *Raptor Res.*, 16, 91-92, 1982.

Allen, R.P. and Peterson, R.T., The hawk migrations at Cape May Point, New Jersey, *Auk*, 53, 393-404, 1936.

Amar, A., Thirgood, S., Pearce-Higgins, J. and Redpath, S., The impact of raptors on the abundance of upland passerines and waders, *Oikos*, 117, 1143-1152, 2008.

Andersson, M. and Norberg, R.Å., Evolution of reversed sexual size dimorphism and role partitioning among predatory birds with a size scaling of flight performance, *Biol. J. Linne. Soc.*, 15, 105-130, 1981.

Austin, G.E., Calbrade, N.A., Mellan, H.J., Musgrove, A.J., Hearn, R.D., Stroud, D.A., Wotton, S.R. and Holt, C.A., Waterbirds in the UK 2012/13: The Wetlands Bird Survey, BTO/RSPB/JNCC, Thetford, 2014.
(http://www.bto.org/volunteer-surveys/webs/publications/webs-annual-report).

Baines, D. and Richardson, M., Hen Harriers on a Scottish moor: multiple factors predict breeding density and productivity, J. Appl, Ecol., 50, 1397-1405, 2013.

Baker, R.R. and Bibby, C.J., Merlin *Falco columbarius* predation and theories of the evolution of bird coloration, *Ibis*, 129, 259-263, 1987.

Ballard, J.W.O. and Whitlock, M.C., The incomplete natural history of mitochondria, *Molecular Ecology*, 13, 729-744, 2004.

Bauer, E.N., Winter roads raptor survey in El Paso County, Colorado, 1962-1979, *Raptor Res.*, 16, 10-13, 1982.

Becker, D.M., Food habits of Richardson's Merlins in southeastern Montana, *Wilson Bull.*, 97, 226-230, 1985.

Becker, D.M. and Sieg, C.H., Breeding chronology and reproductive success of Richardson's Merlins in southeastern Montana, *Raptor Res.*, 19, 54-59, 1985.

Becker, D.M. and Sieg, C.H., Home range and habitat utilization of breeding male Merlins, *Falco columbarius*, in southeastern Montana, *Can. Field-Nat.*, 101, 398-403, 1987a.

Becker, D.M. and Sieg, C.H., Eggshell quality and organochlorine residues in eggs of Merlin *Falco columbarius* in southeastern Montana, *Can. Field-Nat.*, 101, 369-372, 1987b.

Bell, D.A., Griffiths, C.S., Caballero, I.C., Hartley, R.R. and Lawson, R.H., Genetic evidence for global dispersion in the Peregrine Falcon (*Falco peregrinus*) and affinity with the Taita Falco (*Falco fasciinucha*), *J. Raptor Res.*, 48, 44-53, 2014.

Bengtson, S-A., (Hunting methods and prey of an Icelandic population of Merlins (*Falco columbarius*), *Fauna och Flora*, 70, 8-12, 1975.

Bent, A.C., *Life histories of North American Birds of Prey (Part 2)*: Smithsonian Institute, United States National Museum Bulletin 170, US Government Printing Office, Washington, 1938.

Bergman, G., The food of birds of prey and owls in Fenno-Scandia, *Brit. Birds*, 54, 307-320, 1961.

Beukeboom, L., Dijkstra, C., Daan, S. and Meijer, T., Seasonality of clutch size determination in the Kestrel *Falco tinnunculus*: An experimental approach, *Ornis Scand.*, 19, 41-48, 1988.

Bibby, C.J., Merlins in Wales: site occupancy and breeding in relation to vegetation, *J. App. Ecol.*, 23, 1-12, 1986.

Bibby, C.J., Foods of breeding Merlins in Wales, *Bird Study*, 34, 64-70, 1987.

Bibby, C.J. and Nattrass, M., The breeding status of the Merlin in Britain, *Brit. Birds*, 79, 170-185, 1986.

Bond, R., Speed and eyesight of a Pigeon Hawk, *Condor*, 38, 85, 1936.

Boyce Jr., D.A., Merlins and the behaviour of wintering shorebirds, *Raptor Res.*, 19, 95-96, 1985.

Boyle, G.L., Skylark using car as refuge from Merlin, *Brit. Birds*, 84, 18, 1991.

Brady, R., Spotted Sandpiper plunges underwater to avoid Merlin attack, *Passenger Pigeon*, 61, 455-456, 1999.

Brazil, M., *Birds of East Asia, eastern China, Taiwan, Korea, Japan and eastern Russia*, Christopher Helm, London, 2009.

Brewster, W., The Birds of the Lake Umbagog Region of Maine, *Bull. Mus. Comp. Zool.*, 64, 355-362, 1925.

Brodkorb, P., Catalogue of fossil birds: Part 2 (Anseriformes through Galliformes), *Bull. Florida State Mus. Biol. Sci.*, 8, 1964.
Brown, L., *British Birds of Prey*, Collins, London, 1976.

Brown, R.H., The food of certain birds of prey, *Brit. Birds*, 28, 257-258, 1935.

Bruderer, B. and Boldt, A., Flight characteristics of birds: I. radar measurements of speeds, *Ibis*, 143, 178-204, 2001.

Bruinzeel, L.W. and van de Pol, M., Site attachment of floaters predicts success in territory acquisition, *Behav. Ecol.*, 15, 290-296, 2004.

Buchanan, J.B., The effect of kleptoparasitic pressure on hunting behaviour and performance of host Merlins, *J. Raptor Res.*, 22, 63-64, 1988.

Buchanan, J.B., A comparison of behaviour and success rates of Merlins and Peregrine Falcons when hunting Dunlin in two coastal habitats, *J. Raptor Res.*, 30, 93-98, 1996.

Buchanan, J.B., Change in the winter occurrence of Merlins at a western Washington estuary following recovery of Peregrine Falcon populations, *J. Raptor Res.*, 43, 149-151, 2009.

Buchanan, J.B., Is simultaneous hunting in winter by Merlins cooperative?, *J. Raptor Res.*, 44, 156-158, 2010.

Buchanan, J.B., Change in Merlin hunting behaviour following recovery of Peregrine Falcon populations suggests mesopredator suppression, *J. Raptor Res.*, 46, 349-356, 2012.

Buchanan, J.B., Schick, C.T., Brennan, L.A. and Herman, S.G., Merlin predation on wintering Dunlins: hunting success and Dunlin escape tactics, *Wilson Bull.*, 100, 108-118, 1988.

Buden, D.W., The birds of Long Island, Bahamas, *Wilson Bull.*, 104, 220-243, 1992a.

Buden, D.W., The birds of Exumas, Bahama Islands, *Wilson Bull.*, 104, 674-698, 1992b.

Butterfield, A., *Falco columbarius subaesalon* Brehm: A valid race, *Brit. Birds*, 47, 342-347, 1954.

Cade, T.J., *The Falcons of the World*, Cornell University Press, Ithaca and Collins, London, 1982.

Campioni, L., Delgado, M.D.M. and Penteriani, V., Social status influences microhabitat selection: breeder and floater Eagle Owls *Bubo bubo* use different post sites, *Ibis*, 152, 569-579, 2010.

Carere, C., Montanino, S., Moreschini, F., Zoratto, F., Chiarotti, F., Santucci, D. and Alleva, E., Aerial flocking patterns of wintering Starlings, *Sturna vulgaris*, under different predation risk, *Anim. Behav.*, 77, 101-107, 2009.

Chandler, R.B., Strong, A.M. and Kaufman, C.C., Elevated leadlevels in urban House Sparrows: a threat to Sharp-shinned Hawks and Merlins?, *J. Raptor Res.*, 38, 62-68, 2004.

Charpentier, G., Louat, F., Bonmatin, J-M. Marchand, P.A., Vanier, F., Locker, D. and Decoville, M., Lethal and sublethal effects of imidacloprid, after chronic exposure, on the insect model *Drosophila melanogaster*, Environmental Science and Technology (201) 48 (7) 4096-4102 - doi : 10.1021/es405331c, 2014.

Chavez-Ramirez, F., Vose, G.P. and Tennant, A., Correlation between raptor and songbird numbers at a migratory stopover site, *Wilson Bull.*, 106, 150-154, 1994.

Chesser, R.T., Banks, R.C., Barker, F.K., Cicero, C., Dunn, J.L., Kratter, A.W., Lovette, I.J., Rasmussen, P.C., Remsen Jr, J.V., Rising, J.D., Stotz, D.F. and Winker, K., Fifty-third supplement to the American Ornithologists' Union check-list of North American birds, *Auk*, 129, 573-588, 2012.

Clark, A.J. and Scheuhammer, A.M., Lead poisoning in upland-foraging birds of prey of Canada, *Ecotoxicology*, 12, 23-30, 2003.

Clark, W.S., Migration of the Merlin along the coast of New Jersey, *Raptor Res.*,19, 85-93, 1985.

Clarke, R. and Scott, D., Breeding season diet of the Merlin in County Antrim, *Irish Birds*, 5, 205-206, 1994.

Cochrane, W.W. and Applegate, R.D., Speed of flight of Merlins and Peregrine Falcons, *Condor*, 88, 397-398, 1986.

Cooper, J.E., Developmental abnormalities in two British falcons (*Falco* spp.), *Avian Pathology*, 13, 639-645, 1984.

Cooper, J.E. and Forbes, N.A., Studies on morbidity and mortality in the Merlin (*Falco columbarius*), *Veterinary Record*, 118, 232-235, 1986.

Cooper, J.M., Merlin (*Falco columbarius*) preys on flying dragonflies, *British Columbia Birds*, 6, 15-16, 1996.

Corso, A., Raptor migration across the Strait of Messina, southern Italy, *Brit. Birds*, 94, 196-202, 2001.

Cosnette, B.L., Successive use of same site by two female Merlins, *Scot. Birds*, 13, 118, 1984.

Cosnette, B.L., Apparent bigamy in Merlins and co-operation of two females with large young, *North-East Scotland Bird Report 1990,* 74-75, 1991.

Craib, J.K., Merlin follows prey underground, *Scot. Birds*, 17, 236, 1994.

Craighead, J. and Craighead, F., Nesting Pigeon Hawks, *Wilson Bull.*, 52, 241-248, 1940.

Cramp, S. and Simmons, K.E.L. (editors), *Handbook of the Birds of Europe, the Middle East and North Africa: The Birds of the Western Palearctic*, Oxford University Press, 1980.

Cresswell, W., Escape responses by Redshanks, *Tringa totanus*, on attack by avian predators, *Anim. Behav.*, 46, 609-611, 1993.

Cresswell, W., Song as a pursuit-deterrent signal, and its occurrence relative to other anti-predation behaviours of Skylark (*Alauda arvensis*) on attack by Merlins (*Falco columbarius*), *Behav. Ecol. Sociobiol.*, 34, 217-223, 1994.

Cresswell, W., Surprise as a winter hunting strategy in Sparrowhawks (*Accipiter nisus*), Peregrines (*Falco peregrinus*) and Merlins (*F. columbarius*), *Ibis*, 138, 684-692, 1996.

Cresswell, W. and Whitfield, D.P., The effects of raptor predation on wintering wader populations at the Tyninhame estuary, southeast Scotland, *Ibis*, 136, 223-232, 1994.

Crochet, P-A., A taiga Merlin on the Azores: an overlooked vagrant to Europe?, *Birding World*, 21, 114-116, 2008.

Dean, T., Merlin preying on Leach's Petrel, *Brit. Birds*, 81, 395, 1988.

DeGregorio, B.A., Chiavacci, S.J., Weatherhead, P.J., Willson, J.D., Benson, T.J. and Sperry, J.H., Snake predation on North American bird nests: culprits, patterns and future directions, *J. Avian Biol.*, 45, 325-333, 2014.

Dekker, D., Peregrine Falcon and Merlin predation of on small shorebirds and passerines in Alberta, *Can. J. Zool.*, 66, 925-928, 1988.

Dekker, D., Over-ocean flocking by Dunlins, *Calidris alpina*, and the effect of raptor predation at Boundary Bay, British Columbia, *Can. Field-Nat.*, 112, 694-697, 1998.

Dekker, D., Peregrine Falcon predation on Dunlins and ducks and kleptoparasitic interference from Bald Eagles wintering at Boundary Bay, British Columbia, *J. Raptor Res.*, 37, 91-97, 2003.

Dekker, D. and Ydenburg, R., Raptor predation on wintering Dunlins in relation to the tidal cycle, *Condor*, 106, 415-419, 2004.

Dickson, R.C., A Merlin roost in Wigtownshire, *Scot. Birds*, 7, 288-292, 1973.

Dickson, R.C., Habitat preferences and prey of Merlins in winter, *Brit Birds*, 81, 269-274, 1988.

Dickson, R.C., Restricted winter range of a Merlin in west Galloway, *Scot. Birds*, 15, 131-132, 1989.

Dickson, R.C., Hunting times by Merlins in winter, *Scot. Birds*, 17, 56-58, 1993.

Dickson, R.C., Mating times of Merlins, *Scot. Birds*, 17, 160-161, 1994.

Dickson, R.C., Nest reliefs and feeding rates of Merlin, *Scot. Birds*, 18, 20-23, 1995.

Dickson, R.C., The hunting behaviour of Merlins in winter in Galloway, *Scot. Birds*, 18, 165-169, 1996.

Dickson, R.C., Cannibalism in a Merlin brood, *Scot. Birds*, 19, 167, 1998a.

Dickson, R.C., Merlin's sunning behaviour in winter, *Scot. Birds*, 19, 176, 1998b.

Dickson, R.C., Hunting associations between Merlins and Hen Harriers, *Scot. Birds*, 19, 245, 1998c.

Dickson, R.C., Size of Merlin pellets in winter and summer in Galloway, Scot. Birds, 20, 31-33, 1999.

Dickson, R.C., Prey captured and attacked by Merlins in winter, *Scot. Birds*, 21, 116-117, 2000 + erratum *Scot. Birds*, 22, 68, 2001.

Dickson, R.C., Sunning behaviour by a fledgling Merlin, *Scot. Birds*, 24, 43, 2003.

Dickson, R.C., Female Merlin hunting in her nest area, *Scot. Birds*, 25, 54, 2005.

Downes, C.M., Population trends in raptors from the Breeding Bird Survey, *Bird Trends*, 9, 9-12, 2003.

Duncan, K., Merlin killing and retrieving a Dipper from a loch, *Scot. Birds*, 16, 40, 1990.

Duncan, J.R. and Bird, D.M., The influence of relatedness and display effort on the mate choice of captive female American Kestrels, *Anim. Behav.*, 37, 112-117, 1989.
Duquet, M. and Nadal, R., (The capture of bats by raptors), *Ornithos*, 19, 184-195, 2012.

Ellis, D.H., First breeding records of Merlins in Montana, *Condor*, 78, 112-114, 1976.

Ellis, P.M. and Okill, J.D., Breeding ecology of the Merlin *Falco columbarius* in Shetland, *Bird Study*, 37, 101-110, 1990.
Elts, J., Kuresoo, A., Leibak, E., Lilleleht, V., Luigujõe, L., Lõhmus, A., Mägi, E. and Ots, M., (Status and population of Estonian birds, 1998-2002), *Hirundo*, 16, 58-85, 2003.

Elts, J., Leito, A., Leivits, A., Luigujõe, L., Mägi, E., Nellis, R., Ots, M. and Pehlak, H., (Status and population of Estonian birds, 2008-2012), *Hirundo*, 26, 80-112, 2013.

Emslie, S.D., The late Pleistocene (Rancholabrean) avifauna of Little Box Elder Cave, Wyoming, *Univ. Wyoming. Contrib. Geol.*, 23, 63-82, 1985.
Enderson, J.H., Flatten, C. and Jenny, J.P., Peregrine Falcons and Merlins in Sinaloa, Mexico in winter, *J. Raptor Res.*, 25, 123-126, 1991.

Espie, R.H.M., *Factors affectig breeding performance in Merlins (Falco columbarius): Tactics, roles and responses of two sexes*, PhD thesis, University of Saskatchewan, 2000.

Espie, R.H.M., James, P.C., Warkentin, I.G. and Oliphant, L.W., Ecological correlates of molt in Merlins (*Falco columbarius*), *Auk*, 113, 363-369, 1996.

Espie, R.H.M., Oliphant, L.W., James, P.C., Warkentin, I.G. and Lieske, D.J., Age-dependent breeding performance in Merlins (*Falco columbarius*), *Ecology*, 81, 3404-3415, 2000.

Espie, R.H.M., James, P.C., Oliphant, L.W., Warkentin, I.G. and Lieske, D.J., Influence of nest-site and individual quality on breeding performance in Merlins *Falco columbarius*, *Ibis*, 146, 623-631, 2004.

Etheridge, B., Breeding raptors in Scotland: a review, *Scot. Birds*, 33, 38-45, 2013

Ewing, S.R., Rebecca, G.W., Heavisides, A., Court, I.R., Lindley, P., Ruddock, M., Cohen, S. and Eaton, M.A., Breeding status of Merlins *Falco columbarius* in the UK in 2008, *Bird Study*, 58, 379-389, 2011.

Feldsine, J.W. and Oliphant, L.W., Breeding behavior of the Merlin: the courtship period, *Raptor Res.*, 19, 60-67, 1985.

Fernández-Bellon, D. and Lusby, J., The feeding ecology of Merlin *Falco columbarius* during the breeding season in Ireland, and an assessment of current diet analysis methods, *Irish Birds*, 9, 159-164, 2011a.

Fernández-Bellon, D. and Lusby, J., *Irish Merlin Ecological Research, 2011 Report.* Report to the National Parks and Wildlife Service, BirdWatch Ireland, Banagher, 2011b.

Fernández-Bellon, D. and Lusby, J., The effectiveness of playback as a method for monitoring breeding Merlin *Falco columbarius* in Ireland, *Irish Birds*, 9, 155-158, 2011c.

Fiuczynski, K.D. and Sömmer, P., Adaptation of two falcon species *Falco femoralis* and *Falco subbuteo* to an urban environment, in Chancellor, R.D. and Meyburg, B.-U. (eds.), *Raptors at Risk*, WWGBP/Hancock House, 463-467, 2000.
Fleet, D.M., Merlins stalking Dunlins on foot, *Brit. Birds*, 86, 181, 1993.

Forsman, D., *The Raptors of Europe and the Middle East*, T&AD Poyser, London, 1999.

Fox, G.A., Notes on the western race of the Pigeon Hawk, *Blue Jay*, 22, 140-147, 1964.

Fox, G.A., Recent changes in the reproductive success of the Pigeon Hawk, *J. Wildl. Manage.*, 35, 122-128, 1971.
Fox, G.A., A simple method of predicting DDE contamination and reproductive success of populations of DDE-sensitive species, *J. Applied Ecol.*, 16, 737-741, 1979.

Fox, G.A. and Donald, T., Organochlorine pollutants, nest-defence behavior and reproductive success in Merlins, *Condor*, 82, 81-84, 1980.

Fox, G.A., Anderka, F.W., Lewin, V. and MacKay, W.C., Field assessment of eggshell quality by beta-backscatter, *J. Wildl. Manage.*, 39, 528-534, 1975.

Fyfe, R.W., Risebrough, R.W. and Walker, W., Pollutant effects on the reproduction of the Prairie Falcons and Merlins of the Canadian Prairies, *Can. Field-Nat.*, 90, *346-355*, 1976.

Galtier N., Nabholz B., Glémin S. and Hurst G.D., Mitochondrial DNA as a marker of molecular diversity: a reappraisal, *Molecular Ecology*, 18, 4541-4550, 2009.

Garner, M., Identification and vagrancy of American Merlins in Europe, *Birding World*, 15, 468-480, 2002.

Génsbøl, B. and Thiede, W., *Greif Vögel*, BLV, Munich, 2004.

Gjershaug, J.O., Thingstad, P.G., Eldøy, S. and Byrkjeland, S., *Norsk fugleatlas*, pp130-131, Norsk Ornitologisk Forening, Klæbu, 1994.

Gray, A.P., *Bird Hybrids*, Commonwealth Agricultural Bureaux, Farnham Royal, Buckinghamshire, 1958.

Greaves, J.W., Food concealment by Merlins, *Brit. Birds*, 309-311, 1968.

Greenwood, P.J., Mating systems, philopatry and dispersal in birds and mammals, *Anim. Behav.*, 28, 1140-1162, 1980.

Greenwood, J.J.D., Crick, H.Q.P. and Bainbridge, I.P., Numbers and international importance of raptors and owls in Britain and Ireland in *Birds of Prey in a Changing Environment*, D.B.A. Thompson, D.B.A., Redpath, S.M., Fielding, A.H., Marquiss, M. and Galbraith, C.A., Scottish Natural Heritage, The Stationery Office Scotland, Edinburgh, 2003.

Gricks, N., Short-eared Owl preying on Merlin, *Brit. Birds*, 98, 26, 2005.

Griffiths, C.S., Syringeal morphology and the phylogeny of the Falconidae, *Condor*, 96, 127-140, 1994a.

Griffiths, C.S., Monophyly of the Falconiformes based on syringeal morphology, *Auk*, 111, 787-805, 1994b.

Griffiths, C.S., Correlation of functional domains and rates of nucleotide substitution in cytochrome *b.*, *Mol. Phylogenet. Evol.*, 7, 352-365, 1997.

Griffiths, C.S., Phylogeny of the Falconidae inferred from molecular and morphological data, *Auk*, 116-130, 1999.

Groombridge, J.J., Jones, C., Bayes, M.K., van Zyl, A.J., Carillo, J., Nichols, R.A. and Bruford, M.W., A molecular phylogeny of African Kestrels with reference to divergence across the Indian Ocean, *Mol. Phylogenet. Evol.*, 25, 267-277, 2002.

Haak, B.A., Winter occurrence of three Merlin sub-species in southwestern Idaho, USA., *J. Raptor Res.*, 46, 220-223, 2012.

Haak, B.A. and Buchanan J.B., Bathing and drinking behaviour of wintering Merlins, *J. Raptor Res.*, 46, 224-226, 2012.

Hackett, S.J., Kimball, R.T., Reddy, S., Bowie, R.C., Braun, E.L., Chojnowski, J.L., Cox, W.A., Han, K., Harshman, J., Huddleston, C.D., Marks, B.D., Miglia, K.J., Moore, W.S., Sheldon, F.H., Steadman, D.W., Witt, C.C. and Yuri, T., A phylogenomic study of birds reveals their evolutionary history, *Science*, 320, 1763-1768, 2008.

Haffield, P., Merlin in mid and south Wales, *Birds in Wales*, 9, 41-49, 2012.

Hagen, Y., *Rovfuglene og viltpleien*, Gyldendal Norsk Forlag, Oslo 1952.

Hager, S.B., Human-related threats to urban raptors, *J. Raptor Res.*, 43, 210-226, 2009.

Hakkarainen, H. and Korpimäki, E., Competitive and predatory interactions among raptors: an observational and experimental study, *Ecology*, 77, 1134-1142, 1996.

Hakkarainen, H., Korpimäki, E., Huhta, E. and Palokangas, P., Delayed maturation in plumage colour: evidence for the female-mimicry hypothesis in the Kestrel, *Behav. Ecol. Sociobiol.*, 33, 247-251, 1993.

Hallmann, C.A., Foppen, R.P.B. , van Turnhout, C.A.M., de Kroon, H. and Jongejans, E., Declines in insectivorous birds are associated with high neonicotinoid concentrations, *Nature*, 511, 341-342, 17 July 2014.

Hambly, C., Harper, E.J. and Speakman, J.R., The energy cost of loaded flight is substantially lower than expected due to alterations in flight kinematics, *J. Exp. Biol.*, 207, 3969-3976, 2004.
Haug, E., Merlin feeding on road-kills, *Raptor Res.*, 19, 103, 1985.

Heavisides, A., British and Irish Merlin recoveries, 1911 – 1984, *Ringing and Migration*, 8, 29-41, 1987.

Heavisides, A., *Merlin* in Wernham, C., Toms, M., Marchant, J., Clark, J., Siriwardena, G. and Baillie, S. (eds), *The Migration Atlas: Movements of the Birds of Britain and Ireland*, T&AD Poyser, London, 2002 (reprinted 2008), pp250-252.

Hedenström, A. and Rosén, M., Predator versus Prey: on aerial hunting and escape strategies in birds, *Behav. Ecology*, 12, 150-156, 2001.

Heggøy, O. and Øien, I.J., Conservation status of birds of prey and owls in Norway, *Norsk Ornitologisk Forening* Report 1, 2014.

Henny, C.J., Bean, J.R. and Fyfe, R.W., Elevated Heptachlor Epoxide and DDE residues in a Merlin that died after migrating, *Can. Field-Nat.*, 90, 361-363, 1976.

Higgins, R.J. and Hannam, D.A.R., Lymphoid leucosis in a captive Merlin (*Falco columbarius*), *Avian Pathology*, 14, 445-447, 1985.

Hills, I., Strange nesting site of a Merlin *Falco columbarius* in Finland, *Ornis Fennica*, 57, 41, 1980.
Hilty, S.L., *Birds of Venezuela*, Christopher Helm, London, 2002.

Hodson, K.A., *Some aspects of the nesting ecology of Richardson's Merlin (Falco columbarius richardsonii) on the Canadian prairies*, MSc thesis, University of British Columbia, Vancouver, Canada, 1976.

Hodson, K., Prey utilised by Merlins nesting in shortgrass prairies of southern Alberta, *Can. Field-Nat.*, 92, 76-77, 1978.

Hoffman, S.W. and Smith, J.P., Population trends in migratory raptors in western North America, 1977-2001, *Condor*, 105, 397-419, 2003.

Hogstad, O., Improved breeding success of Fieldfares *Turdus pilaris*, nesting close to Merlins *Falco columbarius*, *Fauna Norvegica*, 5, 1-4, 1981.

Holmes, T.L., Knight, R.L., Stegall, L. and Craig, G.R., Responses of wintering grassland raptors to human disturbance, *Wildl. Soc. Bull.*, 21, 461-468, 1993.

Houston, C.S. and Hodson, K.A., Resurgence of breeding Merlins, *Falco columbarius richardsonii*, in Saskatchewan grasslands, *Can. Field-Nat.*, 111, 243-248, 1997.

Hoyt, D.F., Practical methods of estimating volume and fresh weight of bird eggs, *Auk*, 96, 73-77, 1979.

Hull, J.M., Pitzer, S., Fish, A.M., Ernest, H.B. and Hull A.C., Differential migration in five species of raptors in central coastal California, *J. Raptor Res.*, 46, 50-56, 2012.

Hunt, W.G., Raptor floaters at Moffat's equilibrium, *Oikos*, 82, 191-197, 1998.

Ingram, G.C.S., Nesting habits of Merlins in south Glamorganshire, *Brit. Birds*, 13, 202-26, 1920.

Ivanovski, V., The Merlin (*Falco columbarius*) in northern Belarus, *Buteo*, 13, 67-73, 2003.

Ivanovski, V., Birds of Prey of the Belorussian Poozerie, Vitebsk State University, Vitebsk, 2012.

James, P.C., Urban Merlins in Canada, *Brit. Birds*, 81, 274-277, 1988.

James, P.C. and Oliphant, L.W., Extra birds and helpers at the nests of Richardson's Merlin, *Condor*, 88, 533-534, 1986.

James, P.C. and Smith, A.R., Food habits of urban-nesting Merlins, *Can. Field-Nat.*, 101, 592-596, 1987.

James, P.C., Smith, A.R., Oliphant, L.W. and Warkentin, I.G., Northward expansion of the wintering range of Richardson's Merlin, *J. Field Ornithol.*, 58, 112-117, 1987a.

James, P.C., Oliphant, L.W. and Warkentin, I.G., Close inbreeding in the Merlin (*Falco columbarius*), *Wilson Bull.*, 99, 718-719, 1987b.

James, P.C., Warkentin, I.G. and Oliphant, L.W., Turnover and dispersal in urban Merlins *Falco columbarius*, *Ibis*, 131, 426-429, 1989.

Järvinen, O. and Koskimies, P., Dynamics of the status of threatened birds breeding in Finland, 1935-1985, *Ornis Fennica*, 67, 84-97, 1990.

Jenkins, M.A., Opportunistic hunting techniques of the Merlin, *Ontario Bird Banding*, 8, 40-41, 1972.

Johnson, W.J. and Coble, J.A., Notes on the food habits of Pigeon Hawks, *Jack-Pine Warbler*, 45, 97-98, 1967.

Jourdain, F.C.R in Bent, A.C., *Life histories of North American Birds of Prey (Part 2)*: Smithsonian Institute, United States National Museum Bulletin 170, US Government Printing Office, Washington, 1938.

Karyakin, I. and Nikolenko, E., Merlin in the Altai-Sayan region, Russia, *Raptors Conservation*, 17, 98-12, 2009.

Kelly, G.M. and Thorpe, J.P., A communal roost of Peregrine Falcons and other raptors, *Brit. Birds*, 86, 49-52, 1993.

Kenyon, K.W., Hunting strategy of Pigeon Hawks, *Auk*, 59, 443-444, 1942.

Kerlinger, O., Cherry, J.D and Powers, K.D., Records of migrant hawks from the North Atlantic Ocean, *Auk*, 100, 488-490, 1983.

Kermott, L.H., Merlins as nest predators, *Raptor Res.*, 15, 94-95, 1981.

Kerr, I., The Merlins in winter in Northumberland, *Birds of Northumbria 1988*, 91-94, 1989.

Kimball, R.T., Parker, P.G. and Bednarz, J.C., Occurrence and evolution of cooperative breeding among the diurnal raptors (Accipitridae and Falconidae), *Auk*, 120, 717-729, 2003.

Kirk, D.A. and Hyslop, C., Population status and recent trends in Canadian raptors: a review, *Biological Conservation*, 83, 91-118, 1998.

Kjellén, N., Differential timing of autumn migration between sex and age groups in raptors at Falsterbo, Sweden, *Ornis Scand.*, 23, 420-34, 1992.

Klaassen, R.H.G., Hake, M., Strandberg, R., Koks, B.J., Trierweiler, C., Exo, K-M., Bairlen, F. and Alerstam, T., When and where does mortality occur in migratory birds? Direct evidence from long-term satellite tracking of raptors, *J. Anim. Ecol.*, 83, 176-184, 2014.
Knapton, R.W. and Sanderson, C.A., Food and feeding behaviour of subarctic-nesting Merlins, *Falco columbarius*, at Churchill, Manitoba, *Can. Field-Nat.*, 99, 375-377, 1985.

Knox, A.G., Richard Meinertzhagen – a case of fraud examined, *Ibis*, 135, 320-325, 1993.

Korpimäki, E., Reversed size dimorphism in birds of prey, especially Tengmalm's Owl *Aegolius funereus*: a test of the 'starvation hypothesis', *Ornis Scand.*, 17, 326-332, 1986.

Korpimäki, E., Factors promoting polygyny in European birds of prey – a hypothesis, *Oecologia*, 77, 278-285, 1988.

Koryakin, A.S., Diurnal raptors and owls in the Murmansk region, in *Status of Raptor Populations in Eastern Fennoscandia, Proceedings of the Workshop*, Kostomuksha, Karelia, 2005.

Krone, O., Pathology: Endoparasites, in *Raptor Research and Management Techniques*, Bird, D.L. and Bildstein, D.R. (Eds.), Hancock House, Blaine, WA, USA, 318-328, 2007.

Krüger, O., The evolution of reversed sexual size dimorphism in hawks, falcons and owls: a comparative study, *Evolutionary Ecology*, 19, 467-486, 2005.

Laidlaw, T.G., Zoological Notes, *Ann. Scot. Nat. History*, 8, 183, 1893.

Laing, K., Food habits and breeding biology of Merlins in Denali National Park, Alaska, *Raptor Res.*, 19, 42-51, 1985.

Lawrence, L de K., Notes on nesting Pigeon Hawks at Pimisi Bay, Ontario, *Wilson Bull.*, 61, 15-25, 1949.

Lieske, D.J., Oliphant, L.W., James, P.C., Warkentin, I.G. and Espie, R.H.M., Age of breeding in Merlins (*Falco columbarius*), *Auk*, 114, 288-290, 1997.

Lieske, D.J., Warkentin, I.G., James, P.C., Oliphant, L.W. and Espie, R.H.M., Effects of population density on survival in Merlins, *Auk*, 117, 184-193, 2000.

Lima, S.L., Ecological and evolutionary perspectives on escape from predatory attack: a survey of North American birds, *Wilson Bull.* 105, 1-47, 1993.

Lindberg, P., Relationship between the diet of Fennoscandian Peregrines *Falco peregrinus* and organochlorines and mercury in their eggs and feathers, with a comparison to the Gyrfalcon *Falco rusticolus*, PhD thesis, University of Gothenburg, 1983.

Little, B. and Davison, M., Merlins *Falco columbarius* using crow nests in Kielder Forest, Northumberland, *Bird Study*, 39, 13-16, 1992.

Little, B., Davison, M. and Jardine, D., Merlins Falco columbarius in Kielder Forest: influences of habitat on breeding performance, *Forest Ecology and Management*, 79, 147-152, 1995.

Lobkov, E.G., The illegal capturing of Gyrfalcons *Falco rusticolus* L. in Kamchatka goes on unabated: the biggest batch of birds confiscated in 2012, *Far East J. Orn.*, 3, 67-72, 2012.

Lusby, J., Fernández-Bellon, D., Norriss, D. and Lauder, A., Assessing the effectiveness of monitor methods for Merlin *Falco columbarius* Ireland: the pilot Merlin Study 2010, *Irish Birds*, 9, 143-154, 2011.

Maa, T.C., A revised checklist and concise host index of Hippoboscidae (Dipteria), *Pacific Insects Monographs,* 2, 261-299, 1969.

MacIntyre, D., *Wild Life of the Highlands: shooting, fishing, natural history and legend*, Philip Allan, London, 1936.

Marsden, A., Population studies of falcons using microsatellite DNA profiling, PhD thesis, University of Nottingham, 2002.

Martin, A.P., A study of a pair of breeding Peregrine Falcons (*Falco peregrinus peregrinus*) during part of the nesting period (BSc dissertation, University of Durham, 1980), in Ratcliffe, D., *The Peregrine Falcon*, T&AD Poyser, London, 1993.

Masman, D., The annual cycle of the Kestrel, *Falco tinnunculus*, a study of behavioural energetics, PhD thesis, University of Groningen, 1986.

Masman, D., Gordijn, M., Daan, S. and Dijkstra, C., Ecological energetics of the Kestrel: field estimates of energy intake throughout the year, *Ardea*, 74, 24-39, 1986.

Masman, D., Daan, S., and Beldhuis, J.A., Ecological energetics of the Kestrel: daily energy expenditure throughout the year based on the time-energy budget, food intake and doubly labelled water methods, *Ardea*, 76, 64-81, 1988a.

Masman, D., Daan, S., and Dijkstra, C., Time allocation in the Kestrel (*Falco tinnunculus*), and the principle of energy minimization, *J. Anim. Ecol.*, 57, 411-432, 1988b.

Matyukhin, A.V., Zabashta, A.V. and Zabashta, M.V., Louse flies (Hippoboscidae) of Falconiformes and Strigiformes in Palearctic, in Birds of Prey in the Dynamic Environment of the Third Millenium: Status and Prospects, Gavriliuk, M.N. (ed), *Proc. 6th Int. Conf. on Birds of Prey and Owls of North Eurasia, Kryvyi Rib*, 2012.

McElheron, A., *Merlins of the Wicklow Mountains*, Currach Press, Dublin, 2005.

McIlwraith, T., *Birds of Ontario*, A. Lawson, Hamilton, Canada, 1886.

McIntyre, C.L., Withers, J.M. and Gates, N.J., Merlins (*Falco columbarius*) scavenge road-killed Snowshoe Hare (*Lepus americanus*) in interior Alaska, *J. Raptor Res.*, 43, 254, 2009.

McMillan, R.L., Raptor persecution on a large Perthshire estate: a historical study, *Scot. Birds*, 31, 195-205, 2011.

Mebs, T., (Causes and age of death of Peregrines *Falco peregrinus* from German and Finnish ringed bird recoveries), *Die Vögelwarte*, 26, 98-105, 1971.

Mebs, T. and Schmidt, D., *Die Griefvögel Europas, Nordafrikas und Vorderasiens*, Franckh-Kosmos Verlags, Stuttgart, 2006.

Mee, A., An overview of monitoring for raptors in Ireland, *Acrocephalus*, 33, 239-245, 2012.

Meek, E.E., The breeding ecology and decline of the Merlin *Falco columbarius* in Orkney, *Bird Study*, 35, 209-218, 1988.

Meinertzhagen, R., Notes on a small collection of birds made in Iraq in the winter of 1922-23, *Ibis*, 6, 601-625, 1924.

Mikkola, H., *Owls of Europe*, T&AD Poyser, London, 1983.

Milkovský, J., *Cenozoic Birds of the World. Part 1: Europe*, Ninox Press, Prague, 2002.

Millais, J.G., *Game Birds and Shooting-Sketches*, Henry Sotheran, London, 1892.

Millsap, B.A. and Allen G.T., Effects of falconry harvest on wild raptors populations in the United States: theoretical considerations and management recommendations, *Wildlife Soc. Bull.*, 34, 1392-1400, 2006.

Morbey, Y.E., Coppack, T. and Pulido, F., Adaptive hypotheses for protandry in arrival to breeding areas: a review of models and empirical tests, *J. Orn.*, 153 Supplement 1, S207-S215, 2012.

Morozov, V.V., (The biology of the Merlin, *Falco columbarius*, on the eastern Bol'shezemelskaya tundra), *Russ. J. Orn.*, Express-issue 9, 6-7, 1997.

Morozov. V.V., Bragin, E.A. and Ivanovski, V.V., *Merlin*, Vitebsk State University, Vitebsk, 2013. In Russian, with small English Summary.

Morrison, J.L., Terry, M. and Kennedy, P.L., Potential factors influencing nest defense in diurnal North American raptors, *J. Raptor. Res.*, 40. 98-110, 2006.

Morrison, M.L. and Walton, B.J., the laying of replacement clutches by Falconiforms and Strigiforms in North America, *Raptor Res.*, 14, 79-85, 1980.

Mueller, H.C., Mueller, N.S., Berger, D.D., Allez, G., Robichaud, W. and Kaspar, J.L., Age and sex differences in the timing of the fall migration of hawks and falcons, *Wilson Bull.,* 112, 214-224, 2000.

Nattrass, M., Clement, P. and Brown, A., *Status of the Merlin in England in 1992*, English Nature, Peterborough, 1993.

Nechaev, V.A., *Ptitsy Yuzhnykh Kuril'skikh ostrovov* (The birds of South Kuril Isles), Nauka Press, Leningrad, 1969. In Russian.

Nelson, R.W., Some aspects of the breeding behaviour of Peregrine falcons on Langara Island, British Columbia, MSc thesis, University of Calgary, 1970.

Nethersole-Thompson, C. and Nethersole-Thompson, D., Nets site selection by birds*, Brit. Birds*, 37, 108-113, 1943.

Newton, I., *Population Ecology of Raptors*, T&AD Poyser, London, 1979.

Newton, I. and Haas, M.B., Pollutants in Merlin eggs and their effect on breeding, *Brit. Birds*, 81, 258-269, 1988.

Newton, I. and Rothery, P., Estimation and limitation of the numbers of floaters in a Eurasian Sparrowhawk population, *Ibis*, 143, 442-449, 2001.

Newton, I., Meek, E.R and Little, B., Breeding ecology of the Merlin in Northumberland, *Brit. Birds*, 71, 376-398, 1978.

Newton, I., Bogan, J., Meek, E.R and Little, B., Organochlorine compounds and shell-thinning in British Merlins (*Falco columbarius*), *Ibis*, 124, 328-335, 1982.

Newton, I., Dale, L. and Little, B., Trends in organochlorine and mercurial compounds in the eggs of British Merlins *Falco columbarius*, *Bird Study*, 46, 356-362, 1999.

Newton, I., Meek, E.R and Little, B., Breeding season foods of Merlins *Falco columbarius* in Northumbria, *Bird Study*, 31, 49-56, 1984.

Newton, I., Meek, E. and Little, B., Population and breeding of Northumbrian Merlins, *Brit. Birds*, 79, 155-170, 1986.

Newton, I., Robson, J.E. and Yalden, D.W., Decline of the Merlin in the Peak District, *Bird Study*, 28, 225-234, 1981.

Nielsen, O.K., *Population ecology of the Gyrfalcon with comparative notes on the Merlin and the Raven*, PhD thesis, Cornell University, 1986.

Nittinger, F., Haring, E., Pinsker, W., Wink, M. and Gamauf, A., Out of Africa? Phylogenetic relationships between *Falco biarmicus* and the other Hierofalcons (Aves: Falconidae), *J. Zool. Syst. Evol. Res.*, 43, 321-331, 2005.

Nittinger, F., Gamauf, A., Pinsker, W., Wink, M. and Haring, E., Phylogeography and population surcture of the Saker Falcon (*Falco cherrug*) and the influence of hybridisation: mitochondrial and microsatellite data, *Mol. Ecol.*, 16, 1497-1517, 2007.

Noble, D.G. and Elliot, J.E., Levels of contaminants in Canadian raptors, 1966 to 1988: effects and temporal trends, *Can. Field-Nat.*, 104, 222-243, 1990.

Norriss, D.W., Haran, B., Hennigan, J., McElheron, A., McLaughlin, D.J., Swan, V. and Walsh, A., Breeding biology of Merlins *Falco columbarius* in Ireland, 1986-1992, *Irish Birds*, 9, 23-30, 2010.

Nygård, T. and Polder, A., Environmental pollutants in eggs of birds of prey in Norway: current situation and time trends, *Norsk Institutt for Naturforskning (NINA) Report 834*, 2012.

Odin, N., Merlins hunting at sea, *Brit. Birds*, 85, 467, 1992.

Oliphant, L.W., Merlins – the Saskatoon falcons, *Blue Jay*, 32, 140-147, 1974.

Oliphant, L.W. and Haug, E., Productivity, population density and rate of increase of an expanding Merlin population, *Raptor Res.*, 19, 56-59, 1985.

Oliphant, L.W. and McTaggart, S., Prey utilized by urban Merlins, *Can. Field-Nat.*, 91, 190-192, 1977.

Oliphant, L.W. and Tessaro, S.V., Growth rates and food consumption of hand-raised Merlins, *Raptor Res.*, 19, 79-84, 1985.

Oliphant, L.W. and Thompson, W.J.P., Food caching behaviour in Richardson's Merlin, *Can. Field-Nat.*, 90, 364-365, 1976.

Olsen, J., Reversed sexual dimorphism and prey size taken by male and female raptors: a comment on Pande and Dahanukar, *J. Raptor Res.*, 47, 79-81, 2013.

Olsen, P., Marshall, R.C. and Gaal, A., Relationships within the genus Falco: a comparison of the electrophoretic patterns of feather proteins, *Emu*, 89, 193-203, 1989.

Orchel, J., *Forest Merlins in Scotland: their requirements and management*, The Hawk and Owl Trust, London, 1992.

Ottosson, U., Ottval, R., Elmberg, J., Green, M., Gustafsson, R., Haas, F., Holmquist, N., Lindström, Å., Nilsson, L., Svensson, M., Svensson, S. and Tjernberg, M., *Fåglarna is Sverige – antal och förekomst*, Swedish Ornithological Society, Halmstad, 2012.

Page, G. and Whiteacre, D.F., Raptor predation on wintering shorebirds, *Condor*, 77, 73-83, 1975.

Pain, D.J., Sear, J. and Newton, I., Lead concentrations in birds of prey in Britain, *Environmental Pollution*, 87, 173-180, 1995.

Palmer, R.S., *Handbook of North American Birds: Volume 5*, Yale University Press, New Haven, 1988.

Pande, S. and Dahanukar, N., Reversed sexual dimorphism and differential prey delivery in Barn Owls (*Tyto alba*), *J. Raptor Res.*, 46, 184-189, 2012.

Pande, S. and Dahanukar, N., Reversed sexual dimorphism and prey delivery: response to Olsen, *J. Raptor Res.*, 47, 81-82, 2013.

Parker, A., Peregrines at a Welsh coastal eyrie, *Brit. Birds*, 72, 104-114, 1979.

Parmelee, P.W. and Oesch, R.D., Pleistocene and recent faunas from Brynjulfson Caves, Missouri, Illinois State Mus. Rept. Invest No. 25, 1972.
Parr, S.J., Occupation of new conifer plantations by Merlins in Wales, *Bird Study*, 38, 103-111, 1991.

Parr, S.J., Changes in the population size and nest sites of Merlins *Falco columbarius* in Wales between 1970 and 1991, *Bird Study*, 41, 42-47, 1994.

Parslow, J.L.F., Changes in the status among breeding birds in Britain and Ireland, *Brit. Birds*, 60, 2-47 and 97-123, 1967.

Pennycuick, C.J., *Modelling the Flying Bird*, Academic Press, London, 2008.

Pennycuick, C.J., Fast, P.L.F., Ballerstädt, N. and Rattenborg, N., The effect of an external transmitter on the drag coefficient of a bird's body, and hence on migration range, and energy reserves after migration, *J. Orn.*, 153, 633-644, 2012.

Petty, S.J., Anderson, D.I.K., Davison, M., Little, B., Sherratt, C.J and Lambin, X., The decline of Common Kestrels (*Falco tinnunculus*) in a forested area of northern England: the role of predation by Northern Goshawks (*Accipiter gentilis*), *Ibis*, 145, 472-483, 2003.

Petty, S.J., Patterson, I.J., Anderson, D.I.K., Little, B. and Davison, M., Numbers, breeding performance, and diet of the Sparrowhawk *Accipiter nisus* and Merlin *Falco columbarius* in relation to cone crops and seed-eating finches, *Forest Ecology and Management*, 79, 133-146, 1995.

Pétursson, G., Þráinsson, G. and Ólafsson, E., (Rare birds in Iceland 1989), *Bliki*, 11, 31-63, 1992.

Philips, J.R., A review and checklist of the parasitic mites (Acarina) of the Falconiformes and Strigiformes, *J. Raptor Res.*, 34, 210-231, 2000.

Philips, J.R. and Dindal, D.L., Raptor nests as a habitat for invertebrates: a review, *Raptor Res.*, 11, 86-96, 1977.

Picozzi, N., Growth and sex of nestling Merlins in Orkney, *Ibis*, 125, 377-382, 1983.

Pitcher, E., Widener, P. and Martin S.J., Winter food caching by the Merlin (*Falco columbarius richardsonii*), *Raptor Res.*, 13, 39-40, 1979.

Potapov, E. and Sale R., *The Gyrfalcon*, T&AD Poyser, London, 2005.

Pyle, P., First-cycle molts in North American Falconiformes, *J. Raptor Res.*, 39, 378-385, 2005.

Pyle, P., Evolutionary implications of synapomorphic wing-molt sequences among falcons (Falconiformes) and parrots (Psittaciformes), *Condor*, 115, 593-602, 2013.

Rae, S., Density and productivity of ground-nesting Merlins on an island with no indigenous terrestrial predators, *Scot. Birds*, 30, 7-13, 2010.

Raim, A., Cochran, W.W. and Applegate, R.D., Activities of a migrant Merlin during an island stopover, *J. Raptor Res.*, 23, 49-52, 1989.

Ratcliffe, D.A., Decrease in eggshell weight in certain birds of prey, *Nature*, 215, 208-210, 1967.

Ratcliffe, D.A., Changes attributable to pesticides in egg breakage frequency and eggshell thickness in some British birds, *J. Appl. Ecol.*, 7, 67-107, 1970.

Ratcliffe, D., *The Peregrine Falcon*, T&AD Poyser, London, 1993 (the date refers to the second edition of the monograph which was originally published in 1980).

Rebecca, G.W., Merlin dusting at roadside, *Scot. Birds*, 14, 187, 1987.

Rebecca, G.W., Short-eared Owl killing a Merlin on the nest, *North-east Scotland Bird Report 1990*, 73, 1991.

Rebecca, G., Forest nesting Merlin apparently specialising on Barn Swallows. *Scot. Birds*, 24, 46-48, 2004.

Rebecca, G.W., The breeding ecology of the Merlin (*Falco columbarius aesalon*) with particular reference to north-east Scotland and land-usage change, PhD thesis, Open University, Great Britain, 2006.

Rebecca, G., Spatial and habitat-related influences on the breeding performance of Merlins in Britain, *Brit. Birds*, 104, 202-216, 2011.

Rebecca, G.W. and Bainbridge, I.P., The breeding status of the Merlin *Falco columbarius* in Britain in 1993-94, *Bird Study*, 45, 172-187, 1998.

Rebecca, G.W., Cosnette, B.L. and Duncan, A., Two cases of a yearling and an adult Merlin attending the same nest, *Scot. Birds*, 15, 45-46, 1988.

Rebecca, G.W., Cosnette, B.L., Duncan, A., Picozzi, N., and Catt, D.C., Hunting distance of breeding Merlins in Grampian indicated by ringed wader chicks as prey, *Scot. Birds*, 16, 38-39, 1990.

Rebecca, G.W., Cosnette, B.L., Hardey, J.J.C. and Payne, A.G., Status, distribution and breeding biology of the Merlin in north-east Scotland, 1980-1989, *Scot. Birds*, 16, 165-183, 1992.

Rebecca, G.W., Weir, D.N. and Steele, L.D., Bluethroats killed by nesting Merlins in Scotland, *Scot. Birds*, 14, 174, 1987.

Rich, T.D., Beardmore, C.J., Berlanga, H., Blancher, P.J., Bradstreet, M.S.W., Butcher, G.S., Demarest, D.W., Dunn, E.H., Hunter, W.C., Inigo-Elias, E.E., Kennedy, J.A., Martell, A.M., Panjabi, A.O., Pashley, D.N., Rosenberg, K.V., Rustay, C.M., Wendt, J.S. and Will, T.C., *Partners in Flight North American Landbird Conservation Plan*, Cornell Lab of Ornithology, Ithaca, NY, 2004.

Rijnsdorp, A., Daan, S., and Dijkstra, C., Hunting in the Kestrel, *Falco tinnunculus*, and the adaptive significance of daily habits, *Oecologia*, 50, 391-406, 1981.

Risebrough, R.W., Florant, G.L. and Berger, D.D., Organchlorine pollutants in Peregrines and Merlins migrating through Wisconsin, *Can. Field-Nat.*, 84, 247-253, 1970.

Ritchie, T.L., Two mid-Pleistocene avifaunas from Colman, Florida, *Bull. Florida State Mus. Biol. Sci.*, 26, 1980.

Rivera-Milán, F.F., Distribution and abundance of raptors in Puerto Rico, *Wilson Bull.*, 107, 452-462, 1995.

Roberts, J.L. and Green, D., Breeding failure and decline on a north Wales moor, *Bird Study*, 30, 193-200, 1983.

Roberts, J.L. and Jones, M.S., Merlin *Falco columbarius* on a NE moor – breeding ecology (1983-1997) and possible determinants of density, *Welsh Birds*, 2, 88-107, 1999.

Robertson, I.S., The origin of migrant Merlins on Fair Island, *Brit. Birds*, 75, 108-111, 1982.

Rodriguez-Durán, A. and Lewis, A.R., Seasonal predation by Merlin on Sooty Mustached Bats in western Puerto Rico, *Biotropica*, 17, 71-74, 1985.

Rolle, C.J., Merlin's sunning behaviour in summer, *Scot. Birds*, 20, 39, 1999.

Rosenfield, R.N., Grier, J.W. and Fyfe, R.W., Reducing Management and Research Disturbance, in Bird, D.M. and BIldstein, K.L.(eds.), *Raptor Research and Management Techniques*, Inst. for Wildl. Res., National Wildlife Federation, 1987 (republished by Hancock House, 2007).

Rowan, W., Observations on the breeding habits of the Merlin, *Brit. Birds*, 15, 122-129, 194-202, 222-231, 246-253, 1921-22. The article was in four parts: I – General environment; II – incubation; III – Rearing of the young; and IV – The young.

Ruddock, M. and Whitfield, D.P., A review of disturbance distances in selected bird species, *Natural Research (Projects) Ltd for Scottish Natural Heritage*, 2007.

Rudebeck, G., The choice of prey and modes of hunting of predatory birds with special reference to their selective effect, *Oikos*, 3, 200-231, 1951.

RUG/RIJP (The raptor group of the University of Groningen and the Rijkdienst voor de IJsselmeerpolders, Lelystad, Holland), Timing of vole hunting in aerial predators, *Mammal Review*, 12, 169-181, 1982.

Ruttledge, W., Captive breeding of the European Merlin (*Falco columbarius aesalon*), *Raptor Res.*, 19, 68-78, 1985.

Rydzewski, W., Longevity Records VII, *The Ring*, 80, 169-171, 1974.

Sale, R.G., *A Complete Guide to Arctic Wildlife*, Christopher Helm, London, 2006.

Salvin, F.H. and Brodrick, W., *Falconry in the British Isles*, John Van Voorst, London, 1855.

Sauer, J.R., Hines, J.E. Fallon, J.E. Pardieck, K.L. Ziolkowski, Jr., D.J. and Link, W.A., *The North American Breeding Bird Survey, Results and Analysis 1966 - 2011. Version 07.03.2013,* USGS Patuxent Wildlife Research Center, Laurel, MD, 2012.

Schick, C.T., Brennan, L.A., Buchanan, J.B., Finger, M.A., Johnson, T.M. and Herman, S.G., Organochlorine contamination in shorebirds from Washington State and the significance for their falcon predators, *Environ. Monit. Assess.*, 9, 115-131, 1987.

Schmutz, J.K., Fyfe, R.W., Banasch, U. and Armbruster, H., Routes and timing of migration of falcons banded in Canada, *Wilson Bull.*, 103, 44-58, 1991.

Scott, R.E., Merlins associating with roosting Starlings, *Brit. Birds*, 61, 527-528, 1968.

Schönwetter, M., *Handbuch der Oologie*, Akademie-Verlag, Berlin, 1967.

Seeland, H.M., Niemi, G.J., Regal, R.R., Peterson, A. and Lapin, C., Determination of raptor migratory patterns over a large landscape, *J, Raptor Res.*, 46, 283-295, 2012.

Servheen, C., Notes on wintering Merlins in western Montana, *Raptor Res.,* 19, 97-99, 1985.

Shaw, G., Merlin chick killed by an adder, *Scot. Birds*, 17, 162, 1994.

Shore, R.F., Malcolm, H.M., Weinberg, C.L., Turk, A., Horne, J.A., Dale, L., Wyllie, I. and Newton, I., Wildlife and Pollution: 1999/2000 Annual Report, *Joint Nature Conservancy Council Report* 321, 2002.

Shubin, A.O., (Number and distribution of the Merlin (*Falco columbarius*) in some regions of the north-eastern part of the USSR), in *Ornitologiya Vol. 19*, Moscow University Press, 75-80, 1984.

Sieg, C.H. and Becker, D.M., Nest-site habitat selected by Merlins in south-eastern Montana, *Condor*, 92, 688-694, 1990.

Simms, C, Talon-grappling by Merlins, *Bird Study*, 22, 261, 1975.

Slagsvold, T., The Fieldfare *Turdus pilaris* as a key species in the forest bird community, *Fauna norv. Ser. C Cinclus*, 2, 65-69, 1979.

Slagsvold, T. and Sonerud, G.A., Prey size and ingestion rate in raptors: importance for sex roles and reversed sexual size dimorphism, J. Avian Biol., 38, 650-661, 2007.

Smith, A.R., The Merlins of Edmonton, *Alberta Nat.*, 8, 188-191, 1978.

Sodhi, N.S., Attempted polygyny by a Merlin, *Wilson Bull.*, 101, 505-506, 1989.

Sodhi, N.S., Pair copulations, extra-pair copulations, and intra-specific nest intrusions in Merlin, *Condor*, 93, 433-437, 1991a.

Sodhi, N.S., House Sparrow, flock size in relation to proximity of Merlin, *Can. Field-Nat.*, 105, 278-279, 1991b.
Sodhi, N.S., Effect of a nesting predator on concealment behaviour of potential prey species, *Can. Field-Nat.*, 105, 395-396, 1991c.

Sodhi, N.S., Central place foraging and prey preparation by a specialist predator, the Merlin, *J. Field Ornith.,* 63, 71-76, 1992.

Sodhi, N.S., Proximate determinants of foraging effort in breeding male Merlins, *Wilson Bull.*, 105, 68-76, 1993a.

Sodhi, N.S., Correlates of hunting range size in breeding Merlins, *Condor*, 95, 316-321, 1993b.

Sodhi, N.S., in Sodhi, N.S, Oliphant, L.W., James, P.C. and Warkentin, I.G., The Birds of North America Online (A. Poole (ed)), Ithaca, Cornell Laboratory of Ornithology, http://bna.birds.cornell.edu/bna/species/044 (Behavior), 2005.

Sodhi, N.S. and Oliphant, L.W., Hunting ranges and habitat use and selection of urban-breeding Merlins, *Condor*, 94, 743-749, 1992.

Sodhi, N.S. and Oliphant, L.W., Prey selection by urban-breeding Merlins, *Auk*, 110, 727-735, 1993.

Sodhi, N.S. and Oliphant, L.W., in Sodhi, N.S, Oliphant L.W., James, P.C. and Warkentin, I.G., The Birds of North America Online (A. Poole (ed)), Ithaca, Cornell Laboratory of Ornithology, http://bna.birds.cornell.edu/bna/species/044 (Food Habits), 2005.

Sodhi, N.S., Didiuk, A. and Olliphant, L.W., Differences in bird abundance in relation to proximity of Merlin nests, *Can. J. Zool.*, 68, 852-854, 1990.

Sodhi, N.S., James, P.C., Warkentin, I.G. and Oliphant, L.W., Breeding ecology of urban Merlins (*Falco columbarius*), *Can. J. Zool.*, 70, 1477-1483, 1992.

Sodhi, N.S., Warkentin, I.G. and Oliphant, L.W., Hunting techniques and success rates of urban Merlins (*Falco columbarius*), *J. Raptor Res.*, 25, 127-131, 1991a.

Sodhi, N.S., Warkentin, I.G., James, P.C. and Oliphant, L.W., Effects of radiotagging on breeding Merlins, *J. Wildl. Manage.*, 55, 613-616, 1991b.

Sodhi, N.S, Oliphant L.W., James, P.C. and Warkentin, I.G., The Birds of North America Online (A. Poole (ed)), Ithaca, Cornell Laboratory of Ornithology, http://bna.birds.cornell.edu/bna/species/044, 2005.

Solonen, T., Factors affecting the structure of Finnish birds of prey communities, *Ornis Fennica*, 71, 156-169, 1994.

Sonerud, G.A., Steen, R., Løw, L.M., Røed, L.T., Skar, K., Selås, V. and Slagsvold, T., Size-allocation of prey from male to offspring via female: family conflicts, prey selection, and evolution of sexual size dimorphism in raptors, Oecologia, 172, 93-17, 2013.

Sonerud, G.A., Steen, R., Selås, V., Aanonsen, O.M., Aasen, G-H., Fagerland, K.L., Fosså, A., Kristiansen, L., Løw, L.M., Rønning, M.E., Skouen, S.K., Asakskogen, E., Johansen, H.M., Johnsen, J.T., Karlsen, L.I., Nyhus, G.C., Røed, L.T., Skar, K., Sveen, B-A., Tveiten, R. and Slagsvold, T., Evolution of parental roles in provisioning birds: diet determines role asymmetry in raptors, *Behav. Ecol.*, 25, 762-772, 2014.

Sonerud, G.A., Steen, R., Løw, L.M., Røed, L.T., Skar, K., Selås, V. and Slagsvold, T., Evolution of parental roles in raptors: prey type determines role asymmetry in the Eurasian Kestrel, Anim. Behav. 96, 31-38, 2014.

Sperber I., and Sperber, C., Notes on the food consumption of Merlins, *Zool. Bidr. Upps.*, 35, 263-8, 1963.

Spofford, W.R. and Amadon, D., Live prey to young raptors – incidental or adaptive?, *J. Raptor Res.*, 27, 180-184, 1993.

Stishov, M.S., Pirdatko, V.I. and Baranyuk, V.V., *Ptitsy ostrova Vrangelya* (*The birds of the Wrangel Island*), Nauka Press, Novosibirsk, 1991. In Russian.

Stubbert, D., Display of Merlin, *Brit. Birds*, 37, 17-18, 1943.

Stutchbury, B.J., Floater behaviour and territory acquisition in male Purple Martins, *Anim. Behav.*, 42, 435-443, 1991.

Svensson S., Svensson, M. and Tjernberg, M., *Svensk fågelatlas*, Vår Fågelvärld Supplement 31, 126-127, 1999.

Swarth, H.S., Systematic status of some northwestern birds, *Condor*, 37, 199-204, 1935.

Temple, D., Merlins plucking and eating dead young, *Brit. Birds*, 101, 687-688, 2008.

Temple, S.A., Sex and age characteristics of North American Merlins, *Bird Banding*, 43, 191-196. 1972a.

Temple, S.A., Systematics and evolution of the North American Merlins, *Auk*, 89, 325-338, 1972b.

Temple, S.A., Chlorinated hydrocarbon residues and reproductive success in eastern North American Merlins, *Condor*, 74, 105-106, 1972c.

Thiollay, J.-M., (Population sizes of migratory raptors in the western Mediterranean), *Alauda*, 45, 115-121, 1977.

Thomas, B., Merlin feeding on rabbit carrion, *Scot. Birds*, 16, 219-220, 1992.

Tinbergen, L., (Observations on the division of labour in Kestrels (*Falco tinninculus*) during the breeding season), *Ardea*, 29, 63–98, 1940.

Titus, K. and Fuller, M.R., Recent trends in counts of migrant hawks from Northeastern North America, *J. Wildl. Manage.*, 54, 463-470, 1990.

Treleaven, R.B., *Peregrine: the private life of the Peregrine Falcon*, Headline Publications, Penzance, 1977. (Ratcliffe p236)

Trimble, S.A., Habitat management series for unique or endangered species: Report 15 Merlin *Falco columbarius*, *USDI Bureau of Land Management Technical Note* 271, 1975.

Tucker, V.A., The deep fovea, sideways vision ad spiral flight paths in raptors, *J. Exp. Biol.* 203, 3745-3754, 2000.
Tyrberg, T., Pleistocene birds of the Palearctic, *web.telia.com/~u11502098/pleistocene.pdf* (2008).

Tømmeraas, P.J., Merlin *Falco columbarius* breeding in nests of Magpie *Pica pica*, Raven, *Corvus corax* and Rough-legged Buzzard *Buteo lagopus* in northern Norway, *Fauna Norvegica*, C16, 83-85, 1993a.

Tømmeraas, P.J., Golden Eagles *Aquila chrysaetos* killed a Merlin *Falco columbarius*, robbed a Wigeon *Anas penelope* nest and probably hunted Ring Ouzels *Turdus torquatus* in their territories, *Fauna Norvegica*, C16, 85-88, 1993b.

Uttendörfer, O., *Neue ergebnisse über die emährung der Greifvögel und Eulen*, Ulmer, Stuttgart, 1952.

van Bars-Klinkenberg, G. and Wattel, J., Merlin (*Falco columbarius*) from Bahia, Brazil, *Ardea*, 52, 225-226, 1964.

Vartapetov, L.G., *Ptitsy severnoi taigi Zapadno-Sibirskoi ravniny* (*The birds of the northern taiga of the West-Siberian Plain*), Nauka Press, Novosibirsk, 1998. In Russian.

Videler, J.J., *Avian Flight*, Oxford University Press, Oxford, 2005.

Videler, J.J., Vossebelt, G., Gnodde, M. and Groenewegen, A., Indoor flight experiments with trained Kestrels: I. Flight strategies in still air with and without added weight, *J. Exp. Biol.*, 134, 173-183, 1988a.

Videler, J.J., Groenewegen, A., Gnodde, M. and Vossebelt, G., Indoor flight experiments with trained Kestrels: II. The effect of added weight on flapping flight kinematics, *J. Exp. Biol.*, 134, 173-183, 1988b.

Virkkala, R., Annual variation in northern Finnish forest and fen bird assemblage in relation to spatial scale, *Ornis Fennica*, 68, 193-203, 1991.

Volkovskaya, E.A. and Kurdyukov, A.B. (Extraordinary high concentration of birds feeding on voles in Southern Primorie in winter 2001/2002),*Russ. J. Ornithol.,* Express-issue 208, 3-16, 2003.

Wallen, M.S., Talon-locking between Merlin and Peregrine, *Brit. Birds*, 85, 496, 1992.

Warkentin, I.G., Meadow vole predation by Merlin wintering in Saskatoon, Saskatchewan, *Raptor Res.*, 19, 104-105, 1985.

Warkentin, I.G., Factors affecting roost departure and entry by wintering Merlins, *Can. J. Zool.*, 64, 1317-1319, 1986.

Warkentin, I.G. and James, P.C., Nest-site selection by urban Merlins, *Condor*, 90, 734-738, 1988.

Warkentin, I.G. and James, P.C., Winter roost-site selection by urban Merlins (*Falco columbarius*), *J. Raptor Res.*, 24, 5-11, 1990.

Warkentin I.G. and Oliphant, L.W., Observations of winter food caching by the Richardson's Merlin, *Raptor Res.*, 19, 100-101, 1985.

Warkentin, I.G. and Oliphant, L.W., Habitat use and foraging behaviour of urban Merlins (*Falco columbarius*) in winter, *J. Zool. Lond.*, 221, 539-563, 1990.
Warkentin, I.G. and West N.H., Ecological energetics of wintering Merlins *Falco columbarius*, *Phys Zool.*, 63, 308-333, 1990.

Warkentin, I.G., James, P.C. and Oliphant, L.W., Body morphometrics, age structure and partial migration of urban Merlins, Auk, 107, 25-34, 1990.

Warkentin, I.G., James, P.C. and Oliphant, L.W., Influence of site fidelity on mate switching in urban-breeding Merlins (*Falco columbarius*), *Auk*, 108, 294-302, 1991.

Warkentin, I.G., James, P.C. and Oliphant, L.W., Assortative mating in urban-breeding Merlins, *Condor*, 94, 418-426, 1992a.

Warkentin, I.G., James, P.C. and Oliphant, L.W., Use of a plumage criterion for aging female Merlins, *J. Field Ornithol.*, 63, 473-475, 1992b.

Warkentin, I.G., Curzon, A.D., Carter, R.E., Wetton, J.H., James, P.C., Oliphant, L.W. and Parkin, D.T., No evidence for extrapair fertilizations in the Merlin revealed by DNA fingerprinting, *Molecular Ecology*, 3, 229-234, 1994.

Warkentin, I.G., Lieske, D.J., Espie, R.H.M. and James, P.C., Close in-breeding and dispersal in Merlins: further examination, *J. Raptor Res.*, 47, 69-74, 2013.

Weir, D.N., Ecological preferences of Speyside Merlins and relationship with Sparrowhawks, *Scot. Birds*, 33, 218-228, 2013.

Wheeler, B.K., *Raptors of Eastern North America*, Princeton University Press, 2003a.

Wheeler, B.K., *Raptors of Western North America*, Princeton University Press, 2003b.

White, C.M., Cade, T.J. and Enderson, J.H., *Peregrine Falcons of the World*, Lynx, Barcelona, 2013a.

White, C.M., Sonsthagen, S.A., Sage, G.K., Anderson,C. and Talbot, S.L., Genetic relationships among some subspecies of the Peregrine Falcon (*Falco peregrinus* L.), inferred from mitochondrial DNA control-region sequences, *Auk*, 130, 78-87, 2013b.

Wiklund, C.G., Increased breeding success for Merlins *Falco columbarius* nesting among Fieldfares *Turdus pilaris*, *Ibis*, 121, 109-111, 1979.

Wiklund, C.G., Fieldfare (*Turdus pilaris*) breeding success in relation to colony size, nest position and association with Merlins (*Falco columbarius*), *Behav. Ecol. Sociobiol.*, 11, 165-172, 1982.

Wiklund, C.G., The adaptive significance of nest defence by Merlin, *Falco columbarius*, males, *Anim. Behav.*, 40, 244-253, 1990a.

Wiklund, C.G., Offspring protection by Merlin *Falco columbarius* females: the importance of brood size and expected offspring survival for defense of the young, *Behav. Ecol. Sociobiol.*, 26, 217-233, 1990b.

Wiklund, C.G., Nest predation and Life-span: Components of variance in LRS among Merlin females, *Ecology*, 76, 1994-1996, 1995.

Wiklund, C.G., Determinants of dispersal in breeding Merlins (*Falco columbarius*), *Ecology*, 77, 1920-1927, 1996.

Wiklund, C.G., Food as a mechanism of density-dependent regulation of breeding numbers in the Merlin *Falco columbarius*, *Ecology*, 82, 860-867, 2001.

Wiklund, C.G. and Andersson, M., Nest predation selects for colonial breeding among Fieldfares *Turdus pilaris*, *Ibis*, 363-366, 1980.

Wiklund, C.G. and Larsson B.L., The distribution of breeding Merlins *Falco columbarius* in relation to food and nest sites, *Ornis Svecica*, 4, 113-122, 1994.

Williams, I.T. and Parr, S.J., Breeding Merlins *Falco columbarius* in Wales in 1993, *Welsh Birds*, 1, 14-20, 1995.

Williamson, K., The migration of the Icelandic Merlin, *Brit. Birds*, 47, 434-441, 1954.

Wink, M., Seibold, I., Lotfikhah, F. and Bednarek, W., Molecular systematics of Holoarctic raptors (Order Falconiformes) in *Holoarctic Birds of Prey*, Chancellor, R.D., Meyburg, B-U. and Ferrero J.J. (eds), ADENEX/ World Working Group on Birds of Prey and Owls, 29-48, 1998.

Wink. M. and Sauer-Gürth, H., Advances in the molecular systematics of African raptors in *Raptors at Risk*, Chancellor, R.D. and Meyburg, B-U. (eds), World Working Group on Birds of Prey and Owls/Hancock House, 135-147, 2000.

Wink, M., Sauer-Gürth, H., Ellis, D. and Kenward R., Phylogenetic relationships in the Hierofalco Complex (Saker-, Gyr-, Lanner-, Lagger Falcon) in Chancellor, R.D. and Meyburg, B-U. (eds), *Raptors Worldwide*, World Working Group on Birds of Prey and Owls/MME, 499-504, 2004.

Wittenburg, S.R. and Smith, K.G., Lack of differential migration in juvenile Merlins and Northern Harriers during fall migration in the Florida Keys, *Wilson J. Ornithol.,* 121, 841-843, 2009.

Wittenburg, S.R., Lehnen, S.E. and Smith, K., Use of stable isotopes of hydrogen to predict natal origins of juvenile Merlins and Northern Harriers migrating through the Florida Keys, *Condor*, 115, 451-455, 2013.

Woodin, N., Observations on Gyrfalcon (*Falco rusticolus*) breeding near Lake Myvatn, Iceland, 1967, *J. Raptor Res.*, 14, 97-124, 1980.

Woodland, D.J., Jaafar, Z. and Knight, M-L., The "Pursuit Deterrent" Function of Alarm Signals, *The American Naturalist*, 115, 748-753, 1980.

Wright, P.M., *Merlins of the south-east Yorkshire Dales*, Tarnmoor, Skipton, Yorkshire, 2005.

Wright, T.F., Schirtzinger, E.E., Matsumoto, T., Eberhard, J.R., Graves, G.R., Sanchez, J.J., Capelli, S., Müller, H., Scharpegge, G.K., Chambers, G.K. and Fleischer, R.C., A multi-locus molecular phylogeny of the parrots (Psittaciformes): support for a Gondwanan origin during the Cretaceous, *Molecular Biology and Evolution,* 25, 2141-2156, 2008.

Zimin, V.B., Sazonov, S.V., Lapshin, N.V., Artemiev, A.V., Medvedev, N.V., Khokhlova, T.Y. and Yakovleva, M.V., A review of rare diurnal raptor species in Karelia, in *Status of Raptor Populations in Eastern Fennoscandia, Proceedings of the Workshop*, Kostomuksha, Karelia, 2005.

Zuberogoitia, I., Martínez, J.A., González-Oreja, J.A., Calvo, J.E. and Zabala, J., The relationship between brood size and prey selection in a Peregrine Falcon population located in a strategic region on the Western European Flyway, *J. Orn.*, 154, 73-82, 2013.

Opposite

Female Eurasian Merlin at an artifical nest, Scotland.

Index